O. Thomas Bouman
David G. Brand
Editors

Sustainable Forests: Global Challenges and Local Solutions

Pre-publication
REVIEWS,
COMMENTARIES,
EVALUATIONS . . .

"The conference entitled *Sustainable Forests: Global Challenges and Local Solutions* held in Saskatoon, Saskatchewan in May 1995 brought together participants from across Canada and around the world, including government officials, scientists and forestry professionals, environmentalists, and members of First Nations.

Speakers presented their visions and hopes, and the challenges and frustrations in utilization of our forests to meet the economic and social needs of communities, without irreversibly damaging the renewal capacities of the world's forests. . . .

Also presented in this volume are a selection of scientific papers delivered at the conference, in which we see some progress being made in quantifying measures of sustainability. . . . this volume will give the reader an opportunity to listen in on some of the conversations and debates between conference participants and provides a snapshot of progress toward achieving sustainable forests at the turn of the twentieth century."

Dvoralai Wulfsohn,
PhD, PEng
Associate Professor,
Department of Agricultural and
Bioresource Engineering,
University of Saskatchewan

Sustainable Forests:
Global Challenges
and Local Solutions

Sustainable Forests: Global Challenges and Local Solutions

O. Thomas Bouman
David G. Brand
Editors

Food Products Press
An Imprint of
The Haworth Press, Inc.
New York • London

Published by

Food Products Press, 10 Alice Street, Binghamton, NY 13904-1580 USA

Food Products Press is an imprint of the Haworth Press, Inc., 10 Alice Street, Binghamton, NY 13904-1580 USA.

Sustainable Forests: Global Challenges and Local Solutions has also been published as *Journal of Sustainable Forestry,* Volume 4, Numbers 3/4 and Volume 5, Numbers 1/2 1997.

Library of Congress Cataloging-in-Publication Data

Sustainable forests: global challenges and local solutions/O. Thomas Bouman, David G. Brand, editors.
     p. cm.
     Papers presented at an international conference held in Saskatchewan, May 1995.
     "Has also been published as Journal of sustainable forestry, volume 4, numbers 3/4 and volume 5, numbers 1/2 1997"–T.p. verso.
     Includes bibliographical references and index.
     ISBN 1-56022-055-4 (alk. paper).–ISBN 1-56022-058-9 (pbk.: alk. paper)
     1. Sustainable forestry–Congresses. 2. Forest management–Congresses. 3. Forest ecology–Congresses. I. Bouman, O. Thomas. II. Brand, David George, 1956- .
SD387.S87S855 1997
33.75–dc21
          96-52446
          CIP

INDEXING & ABSTRACTING

Contributions to this publication are selectively indexed or abstracted in print, electronic, online, or CD-ROM version(s) of the reference tools and information services listed below. This list is current as of the copyright date of this publication. See the end of this section for additional notes.

- *Abstract Bulletin of the Institute of Paper Science and Technology,* Institute of Paper Science and Technology, Inc., 575 14th Street, N.W., Atlanta, GA 30318

- *Abstracts in Anthropology,* Baywood Publishing Company, 26 Austin Avenue, P.O. Box 337, Amityville, NY 11701

- *Abstracts on Rural Development in the Tropics (RURAL),* Royal Tropical Institute (KIT), 63 Mauritskade, 1092 AD Amsterdam, The Netherlands

- *AGRICOLA Database,* National Agricultural Library, 10301 Baltimore Boulevard, Room 002, Beltsville, MD 20705

- *Biostatistica,* Executive Sciences Institute, 1005 Mississippi Avenue, Davenport, IA 52803

- *CNPIEC Reference Guide: Chinese National Directory of Foreign Periodicals,* P.O. Box 88, Beijing, People's Republic of China

- *Environment Abstracts,* Congressional Information Service, Inc., 4520 East-West Highway, Suite 800, Bethesda, MD 20814-3389

- *Environmental Periodicals Bibliography (EPB),* International Academy at Santa Barbara, 800 Garden Street, Suite D, Santa Barbara, CA 93101

- *Forestry Abstracts; Forest Products Abstracts (CAB Abstracts),* c/o CAB International/CAB ACCESS . . . available in print, diskettes updated weekly, and on INTERNET. Providing full bibliographic listings, author affiliation, augmented keyword searching. CAB International, P.O. Box 100, Wallingford, Oxon OX10 8DE, United Kingdom

(continued)

- *GEO Abstracts (GEO Abstracts/GEOBASE),* Elsevier/GEO Abstracts, Regency House, 34 Duke Street, Norwich NR3 3AP, England

- *Human Resources Abstracts (HRA),* Sage Publications, Inc., 2455 Teller Road, Newbury Park, CA 91320

- *INTERNET ACCESS (& additional networks) Bulletin Board for Libraries ("BUBL"), coverage of information resources on INTERNET, JANET, and other networks.*
  - JANET X.29: UK.AC.BATH.BUBL or 00006012101300
  - TELNET: BUBL.BATH.AC.UK or 138.38.32.45 login 'bubl'
  - Gopher: BUBL.BATH.AC.UK (138.32.32.45). Port 7070
  - World Wide Web: http://www.bubl.bath.ac.uk./BUBL/home.html
  - NISSWAIS: telnetniss.ac.uk (for the NISS gateway)

  The Andersonian Library, Curran Building, 101 St. James Road, Glasgow G4 ONS, Scotland

- *Journal of Planning Literature,* Ohio State University, Department of City & Regional Planning, 190 West 17th Avenue, Columbus, OH 43210

- *Referativnyi Zhurnal (Abstracts Journal of the Institute of Scientific Information of the Republic of Russia),* The Institute of Scientific Information, Baltijskaja ul., 14, Moscow A-219, Republic of Russia

- *Sage Public Administration Abstracts (SPAA),* Sage Publications, Inc., 2455 Teller Road, Newbury Park, CA 91320

- *Sage Urban Studies Abstracts (SUSA),* Sage Publications, Inc., 2455 Teller Road, Newbury Park, CA 91320

- *Wildlife Review/Fisheries Review,* U.S. Fish and Wildlife Service, 1201 Oak Ridge Drive, Suite 200, Fort Collins, CO 80525-5589

(continued)

SPECIAL BIBLIOGRAPHIC NOTES

related to special journal issues (separates)
and indexing/abstracting

☐ indexing/abstracting services in this list will also cover material in any "separate" that is co-published simultaneously with Haworth's special thematic journal issue or DocuSerial. Indexing/abstracting usually covers material at the article/chapter level.

☐ monographic co-editions are intended for either non-subscribers or libraries which intend to purchase a second copy for their circulating collections.

☐ monographic co-editions are reported to all jobbers/wholesalers/approval plans. The source journal is listed as the "series" to assist the prevention of duplicate purchasing in the same manner utilized for books-in-series.

☐ to facilitate user/access services all indexing/abstracting services are encouraged to utilize the co-indexing entry note indicated at the bottom of the first page of each article/chapter/contribution.

☐ this is intended to assist a library user of any reference tool (whether print, electronic, online, or CD-ROM) to locate the monographic version if the library has purchased this version but not a subscription to the source journal.

☐ individual articles/chapters in any Haworth publication are also available through the Haworth Document Delivery Services (HDDS).

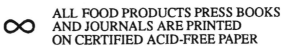

Sustainable Forests: Global Challenges and Local Solutions

CONTENTS

PART TWO: LOCAL SOLUTIONS

ABOUT THE EDITORS

O. Thomas Bouman, PhD, is a forestry graduate from the University of Göttingen, Germany, where he earned his doctorate in 1991. He joined Canada's Model Forest Program in 1993 to work for an association of government agencies, industry, and indigenous people in Saskatchewan. His publications and extension efforts have contributed to the ecological and social foundations of sustainable development in agriculture and forestry.

David G. Brand, PhD, is manager of sustainable development for the State Forests of New South Wales, Australia. Previously, he worked as a scientist at the Petwawa National Forestry Institute and then moved to the Headquarters of the Canadian Forest Service in Ottawa to direct science and sustainable development, including Cananda's Model Forest Program. He has published numerous articles on forestry subjects and chaired a series of national and international forestry-related committees.

Preface

Primary forests that have been created by natural processes are disappearing worldwide at alarming rates chiefly to satisfy the needs of humanity. We have also come to recognize that, regardless of whether they are the product of nature's resilience or human ingenuity, secondary forests do not guarantee conservation of all the complex characteristics of natural forests.

Although this wisdom has been experienced the hard way by many communities on earth for more than 2000 years, it has only recently entered the collective consciousness of the globally acting citizenry. Largely because of environmental advocacy, more and more governments are making use of the notion of sustainable development of the environment including forests.

The most widespread interpretation of sustainable development is probably that of inter-generational equity. In terms of forestry this means that primary forests should no longer be liquidated merely to spur economic growth for a few generations. Instead, the forests created by nature and the processes that regenerate them must be conserved so that the unknown and therefore unique ecological complexity and functionality can contribute to ensuring long-term survival for humanity. Such a precautionary notion of forest sustainability places an unprecedented responsibility on foresters and nations with primary forests that will also be a very ambiguous one. Although humanity is becoming increasingly aware of the absolute scarcity of primary forests, foresters and forestry nations are only economically rewarded when they efficiently supply forest products to solve problems of relative scarcity in a neoclassical sense.

[Haworth co-indexing entry note]: "Preface." Bouman, O. Thomas. Co-published simultaneously in *Journal of Sustainable Forestry* (Food Products Press, an imprint of The Haworth Press, Inc.) Vol. 4, No. 3/4, 1997, pp. xv-xvi; and: *Sustainable Forests: Global Challenges and Local Solutions* (ed: O. Thomas Bouman, and David G. Brand) Food Products Press, an imprint of The Haworth Press, Inc., 1997, pp. xv-xvi. Single or multiple copies of this article are available for a fee from The Haworth Document Delivery Service [1-800-342-9678, 9:00 a.m. - 5:00 p.m. (EST). E-mail address: getinfo@haworth.com].

This volume intends to elucidate the notion of forest sustainability in operational, political and scientific terms. It brings together the ideas, facts and views from Africa, Asia, Europe and the Americas presented at the conference *Sustainable Forests: Global Challenges and Local Solutions*. The presenters accomplished both a reality check and vision building exercise towards forest sustainability. Clearly, there are distinct factors that cause and modify the loss of primary forests across regions. However, the conference has brought home that the involvement and empowerment of communities and non-governmental groups are increasingly seen as the means of changing the way forests are being managed. Nevertheless, the views expressed and the factual evidence presented at the conference have left little doubt that the future of natural forests is bleak. This concern has been further aggravated because, even where the political commitment has been made, operational and scientific support for achieving forest sustainability is generally in its infancy.

O. Thomas Bouman

Acknowledgment

The conference *Sustainable Forests: Global Challenges and Local Solutions* was realized with the support and cooperation of the Canadian Forest Service; Canadian Institute of Forestry; Canadian International Development Agency; Canadian Parks Service; Department of Foreign Affairs and International Trade Canada; Federation of Saskatchewan Indian Nations; International Model Forest Network; Montreal Lake Cree Nation; National Aboriginal Forestry Association (Canada); Prince Albert Grand Council; Saskatchewan Environment and Resource Management; and Weyerhaeuser Canada. It would have been impossible to prepare for this project without the foresight of Gene Kimbley who chaired the local organizing committee joined by Roman Orynik, Kim Clark, Leane Martin, John Doornbos and Dwayne Dye. The dedication of model forest staff and conference volunteers was critical to making the event happen. Special recognition must go to Ian Monteith for arranging the logistics of the conference and assisting in the compilation of this volume.

xvii

INTRODUCTION

Editorial Summary

O. Thomas Bouman
David G. Brand

In late May 1995 the Prince Albert Model Forest Association held an international conference on *Sustainable Forests: Global Challenges and Local Solutions* in Saskatchewan. The conference attracted over three hundred people from the region, Canada, the Americas, Asia, Africa and Europe. A total of 68 papers were presented either orally or as posters. The 24 papers published in this book show the range of issues and subject matter covered. The four day event also included field tours to visit operations, research sites and a Cree community. A six member panel discussion with a focus on Canada concluded the conference. The transcripts have been included in this volume.

Although the notion of sustainable forestry has a long history, it has not been pervasive enough to halt depreciation, degradation and decline of forests, particularly the world's remaining primary forests. This is why the conference organizers intended to attract a

---

O. Thomas Bouman is a forestry graduate from the University of Göttingen, Germany. He is presently Project Administrator of the Prince Albert Model Forest Association Inc., Prince Albert, Sakatchewan, Canada.

David G. Brand is a forestry graduate from the University of British Columbia. He is presently Manager, Sustainable Forest Management, State Forestry of New South Wales, Australia.

The summary refers to authors in this volume unless a publication date is being cited.

[Haworth co-indexing entry note]: "Editorial Summary." Bouman, O. Thomas, and David G. Brand. Co-published simultaneously in *Journal of Sustainable Forestry* (Food Products Press, an imprint of The Haworth Press, Inc.) Vol. 4, No. 3/4, 1997, pp. 3-10; and: *Sustainable Forests: Global Challenges and Local Solutions* (ed: O. Thomas Bouman, and David G. Brand) Food Products Press, an imprint of The Haworth Press, Inc., 1997, pp. 3-10. Single or multiple copies of this article are available for a fee from The Haworth Document Delivery Service [1-800-342-9678, 9:00 a.m. - 5:00 p.m. (EST). E-mail address: getinfo@haworth.com].

broad range of people who would stimulate exchange of ideas beyond the traditional rounds of forestry meetings. The traditionally closed ranks of forest administration, industry and research must open up to make the notion of forest sustainability more effective. This is the chief message that underlies this summary.

We would also like to refer the more deeply interested reader to the results of recent events organized by the Conference on Security and Cooperation in Europe in Montreal (Anon. 1993), by the European Ministerial Process (Anon. 1994) and by the FAO/CFS in Anchorage (Brand 1995).

BACKGROUND

The Canadian Council of Forest Ministers devised a national forest strategy in 1990 that included a new way of delivering forestry programs. The strategy called for partnerships that would hold a broad range of private and public interests in clearly defined forest areas distributed across Canada. Since 1992, a network of ten such model forests has been established at operational scales (0.1 to 1.5 million ha). The Canadian sites have also become part of an international network with a growing number of sites in Mexico, Russia and the United States. More detail about Canada's Model Forest Program and the Prince Albert Model Forest can be found in Brand et al. (1996) and Bouman et al. (1996), respectively.

Semi-annual meetings of all sites have been organized since 1992 to provide for the identification of networking opportunities and review of operational progress towards forest sustainability. It was in fall 1993 when the network agreed to have its spring '95 meeting organized by the Prince Albert Model Forest Association in Saskatchewan. Shortly thereafter the Association decided to organize an international conference on forest sustainability in conjunction with the up-coming network meeting. It was felt that an open invitation to all who know and care about forests could give the newly established model forest partnerships both conceptual guidance and recognition.

In order to put the conference idea into action a series of meetings and discussions were necessary to yield fruitful ideas and to secure logistical support. In this regard, important feedback to the initial conference proposal was provided from within the local partnership, its Research Advisory Committee, the Model Forest Network, and the Canadian Forest Service. Also, the response of colleagues and friends during the joint 1994 convention of American and Canadian foresters in Anchorage, the meeting of the International Boreal Forest Research Association in Saskatoon, a Prince Albert Model Forest lecture tour through Germany, and the First Nations conference Focus on Resources, in Calgary, provided most valuable input.

It soon became quite evident to us that there was no need for another specialized technical or scientific event. Instead, we saw an opportunity to fill a networking gap by organizing a multi-faceted meeting. There was a growing consensus to bring together the diverse political, operational and scientific points of view held by people who were struggling with the notion of forest sustainability in very practical terms. However, the theme of the conference was chosen to reflect that the very challenging extent and pace at which primary forests disappear must allow for participation of all those who are most directly affected by the way forestry operates. We simply extended Chambers' (1983) notion of rural development that the place to start any project is by assessing just what it is that rural project beneficiaries know and want.

ARGUMENTS AND EVIDENCE

It has been credibly argued by many presenters, including the Hon. Anne McLellan for Canada and Janna Kumi for British Columbia, that a significant shift in the development of forest policies has taken place to improve forestry operations and to mitigate their adverse ecological, economic and social effects. Thus forest sustainability is no longer an entirely elusive concept from a policy point of view. However, the emphasis that policy changes should place on the environmental and human welfare dimension of forest sustainability differs between but also within regions.

The European perspectives as presented by Heering and Rykowski are primarily concerned with the environmental dimension. Salleh and Spencer confirm that environmental concerns expressed by Non-Governmental Organizations (NGOs) have become the most important policy drivers for changing private and public forestry operations in Malaysia and Canada, respectively. Lal from India is, on the other hand, most concerned that forestry first and foremost meets the needs of local people whose basic livelihood depends on forests. He is very clear that community empowerment is not a means to an end (e.g., environmental protection) but an end in itself.

Khalikane with his African perspective then takes the discussion of policies and institutional arrangements a step further by demanding respect for indigenous knowledge and cultures. He infers that neocolonial terms of trade have led to an erosion of regional resources and widespread political instability which in turn often leave African peoples with insufficient means to use their forests sustainably.

The First Nations people of North America as represented by Story and Lickers and by Diabo are most adamant about paying respect to indigenous cultures and their inherent management capabilities. At the conference itself, the audience also witnessed debate over forest management responsibilities between First Nations leaders and the Saskatchewan Minister of Environment and Resource Management. The legitimacy of such debate has been corroborated by non-indigenous, international observers such as Fähser from Germany and Plumley from the U.S. who unequivocally pointed out that western societies lack much of their indigenous peoples' relationships with the forest.

Finally, Ostergren and Hollenhorst seem to point to Siberia as being one of the last open forestry frontiers left on this planet. The detailed account of a post perestroika forestry operation suggests that unsustainable use of an abundant but ecologically highly sensitive forest is hampered by classical development problems, e.g., lack of infrastructure and capital.

Apart from the social arguments there is a need to distinguish separate motives that have led to unsustainable forest use. For example, forests in Africa and India are often overused for subsistence or to meet very basic economic needs of people. In North America, forests have been overused to satisfy the needs of affluent markets and profit-oriented corporations. This economic distinction underscores the need for regionally adapted policy solutions.

Simply for reasons of economic power, a leading nation like

Canada would be expected to respond more readily to environmental and social concerns that are related to the disappearance of primary forests. The conference confirmed that field demonstrations of forestry operations provide an excellent opportunity to critically examine and to some extent verify the responsiveness of the forest industry. However, Smith as the locally responsible forester and representative of the Canadian Pulp and Paper Association indicated that it is the international market which exerts an important influence on the operational decisions of the Canadian forest industry.

In addition, Dore et al., Kulshreshtha and Loewen argue in their respective papers that just looking at the exchange value might lead to serious overexploitation of forests. Somewhat similar to the classical forestry debate of the 19th century on whether the land or the forest stands are the assets of a forest enterprise, we now discuss considering forest use values prior to any harvesting decision. In any case, the forest ecosystem produces a host of goods and services for private and public interests. The central question that would warrant further investigation is how does clearcutting affect the production balance of the forest. Does clearcutting represent a shift from public to private production of goods and services? The transcripts of the panel discussion point to this issue.

In terms of environmental concerns the speakers felt that it was important to reiterate the well established fact of rapid worldwide forest decline. Heering estimates the annual global deforestation rate at about 15,000,000 ha. Khasa and Dancik and Salleh point out that the fact that most of the decline affecting primary forests becomes even more dramatic if we include biodiversity considerations into our assessment.

In fact, the three papers presented by Parresol and McCollum, Silbaugh and Betters, and by Bayne and Hobson remind us that we are still at the beginning of quantifying the effects of forestry on spatial and biological diversity of forest ecosystems. Yet, the obvious need for developing manipulative techniques as presented by Reddy et al., and Edson et al. to support reafforestation and biological conservation may be indicative of the ecological urgency to counteract further decline of primary forests and to rehabilitate secondary forests.

THE WAY FORWARD

Four distinct approaches are being offered in this volume for devising solutions that would direct forest users and societies towards improved ecological, economic and social sustainability.

First, changes to the modus operandi must take effect according to Fähser and Locke. Kumi expects, and Flamm shows, that harvest levels must also be reduced in North America. Although Kimmins would object to using clearcutting everywhere, to him ecologically based forestry does not always mean soft footprints in the woods. Action programs for tree planting and rehabilitation of secondary growth, and improving the integration of different land use forms have been recommended by Heering and by Lal.

Secondly, increasing the overall financial commitment to forestry and conservation is an important measure proposed by Heering and by Khalikane to counteract tropical forest decline. Capital investments may also solve some forestry problems in Siberia according to Ostergren and Hollenhorst.

Thirdly, increasing local participation in forest management and research decisions is proposed in a number of cases, e.g., by Schabel for Nicaragua. Differences in the commitment to devolving actual management authority to forest-based communities may be best captured by distinguishing local versus localized decision-making. The latter might more effectively incorporate and recognize the important role of NGOs in ensuring protection and sustainable use of forests. However, the call for increasing local awareness, involvement and participation has been the most commonly held approach at the conference to solving the sustainability impasse.

Fourthly, more reflection on values is probably the most effectively delivered message in this book. Reflection on values can take two directions. Stevenson et al., Maser, and Hummel and Sizykh have taken a science-based approach to value reflection in their papers while Diabo, and Plumley recognize the ethnocentric element. Quantitative survey results presented by Kulshreshtha also suggest there should be, to some extent, culturally based forestry and nature conservation in Saskatchewan. However, we note convergence in principle in that both directions give ecological carrying capacity primacy over economic and social values when there is no harmonic overlap between the three spheres of sustainability.

Some speakers implied that these solutions can be combined. For example, Fähser's comments point to a fundamental shift in operations from manipulative action based on quantitative assumptions to precautionary forestry that is based on the qualitative assumption that humans can not predict the outcome of their forest use. His proposal that operations should follow nature rather than directing it would tie forest management to traditional indigenous thinking.

As for combining participatory and financial solutions, Khalikane sees a need for donor money reaching local people more effectively. He reassures us that local communities are equipped to rationalize investments of money; and they will not divide it amongst their families but instead they will plough it into useful community services.

CLOSING REMARKS

More than 1,200 years ago Pope Gregory II called on the Anglo-Saxon missionary Boniface to strengthen the church's position in what soon became the Holy Roman Empire. History tells us that Boniface axed at some point an oak tree because it was of cultural value to the particularly tenacious indigenous Friesian tribe in northern Germany. This symbolic act was followed first by a loss of indigenous cultures and secondly by a complete devastation of natural primary forests in Germany. In fact, it took almost a millennium before even secondary forests could vigorously regain significant ground. The historical lesson is that cultural conquest can precede forest decline.

We may assume that the global concern about the loss of cultural diversity expressed in this volume might herald the final destiny of the world's primary forests. However, the resilience and survival of some indigenous cultures should provide some vision for perseverance.

Clearly, numerous statements throughout this volume underscore that the discussion about values (including respect for culture, gender and nature) has markedly gained ground. In fact, it may become critical to ensuring better compliance with sustainability principles in forestry. This achievement has largely been made possible through a policy shift to a more participatory approach to forestry as it has become manifest in Canada's Model Forest Program. In this sense, the model forest program exceeds the scope of a forestry action plan.

In summary this volume provides ample evidence that two important principles of sustainability have been adopted by a broad cross section of society—the precautionary and the subsidiarity principle (Dommen 1993).

REFERENCES

Anon. 1993. Proceedings. Seminar of experts on the sustainable development of boreal and temperate forests. Conference on Security and Cooperation in Europe, Montreal, 27 Sept.-1 Oct. 1993. Natural Resources Canada, Canadian Forest Service: Ottawa, Canada.

Anon. 1994. European criteria and most suitable quantitative indicators for sustainable forest management. Adopted by the first expert level follow-up meeting of the Helsinki Conference, Geneva, Switzerland, June 24-25, 1994.

Bouman, O. T., G. Langen and C. T. Bouman. 1996. Sustainable use of the boreal Prince Albert Model Forest in Saskatchewan. The Forestry Chronicle 72, No. 1.

Brand, D. G. (ed.) 1995. Forestry Sector Planning Process conference held Sept 11-15, 1994, Anchorage, Alaska, Canadian Forest Service, F.A.O., Ottawa, Canada.

Brand, D. G., O. T. Bouman, L. Bouthillier, W. Kessler and L. Lapierre. 1996. The Model Forest concept: a model for future forest management? Environment Reviews 4(i) in press.

Chambers, R. 1983. Rural Development: Putting the Last First. Longman, London, UK.

Dommen, E. (ed) 1993. Fair Principles for Sustainable Development: Essays on Environmental Policy and Developing Countries. Edward Elgar Publishing Ltd., Brookfield, VT. p. 170.

Formal Address

Hon. Anne McLellan

Thank you for your kind introduction Gene (Kimbley) and good morning, ladies and gentlemen. It is a pleasure to be here this morning and to bring greetings from the Right Honourable Jean Chrétien, Prime Minister of Canada and the Government of Canada.

I welcome all of you to Saskatoon to this international conference on sustainable forestry, and in particular, I want to welcome those of you from other countries. Please enjoy our country's wonderful spring weather. The number of countries and wide variety of interests represented here this week demonstrate the new and innovative partnerships that we are forging in national and international forest management.

This morning, I would like to discuss a number of important issues facing the forestry community:

- First, the progress we have made to meet global challenges as reported at the Rome Forestry Summit, and at the recent United Nations Commission on Sustainable Development in New York.
- Next, the leading role that Canada has been playing to develop consensus on criteria and indicators, and certification standards for sustainable forest management.

Hon. Anne McLellan is Minister of Natural Resources, Canada.

This address is the opening address given to an international audience at the conference *Sustainable Forests: Global Challenges and Local Solutions*, May 29, 1995, Saskatoon, Saskatchewan, Canada.

[Haworth co-indexing entry note]: "Formal Address." McLellan, Hon. Anne. Co-published simultaneously in *Journal of Sustainable Forestry* (Food Products Press, an imprint of The Haworth Press, Inc.) Vol. 4, No. 3/4, 1997, pp. 11-17; and: *Sustainable Forests: Global Challenges and Local Solutions* (ed: O. Thomas Bouman, and David G. Brand) Food Products Press, an imprint of The Haworth Press, Inc., 1997, pp. 11-17. Single or multiple copies of this article are available for a fee from The Haworth Document Delivery Service [1-800-342-9678, 9:00 a.m. - 5:00 p.m. (EST). E-mail address: getinfo@haworth.com].

11

- And finally, the leadership role that Canada has been playing through the Canadian Model Forest Program, and the International Model Forest Program.

FEDERAL COMMITMENT TO FORESTRY

Let me begin by reiterating that the Government of Canada recognizes the importance of the forest industry. We acknowledge its continuing contribution to economic growth, jobs and trade, and we acknowledge the challenges it faces.

One out of 16 jobs and 350 communities are highly dependent on the forestry sector.

We are proud of our forest industry and we believe its present and future prospects are excellent.

ROME FORESTRY SUMMIT AND UNCSD

In March of this year, I attended the first meeting of Forest Ministers, (FAO) in Rome, to participate in the global forest dialogue. Canada was one of more than 160 countries meeting in Rome at this Forest Summit. We reached consensus on our intention to conserve, manage and use forests in a sustainable manner. Given the number of countries involved, I think you can appreciate how unique an achievement it was to reach this consensus!

Canada believes that the ultimate goal for the world forest community is an international convention that all countries can agree to–which will provide the global blueprint for sustainably managed forests.

I delivered the following messages on behalf of the Government of Canada in Rome:

- First the United Nations has an important role to play in assisting in the development of a global framework for criteria and indicators.
- Second, there is a need to engage those countries and regions not yet involved in formulating criteria and indicators, to become active in developing a global framework for criteria and indicators.
- Third, there is a need for greater international consensus on voluntary, non-legislated schemes for certification of forest products from sustainably managed forests.

- Fourth, there is a need to provide reliable and timely information on the state of forests to help guide policy at the national, regional and global levels.
- And finally, that the complexity of forest issues requires *shared leadership* and partnerships.

The Rome Forest Summit was the preparatory event for the first discussion of forest issues by the United Nations Commission on Sustainable Development in April in New York. The UNCSD meeting assessed the progress made in sustainable forest management since the Rio Earth Summit in 1992.

An important outcome of the New York meeting was the creation of an intergovernmental panel on forests. This panel's tasks include developing recommendations to reconcile the diverging interests and pressures facing forest countries, examining ways to better understand the many functions performed by all types of forests, and working toward consensus on criteria and indicators.

CRITERIA AND INDICATORS

Since the 1992 Earth Summit, dramatic progress has been made in developing criteria and indicators of sustainable forest management. Four key exercises have been instrumental in their progress:

- The International Tropical Timber Organization for tropical forest countries;
- The Helsinki Process for European forest countries;
- The Amazon Pact for Amazonian forests countries; and
- The Montreal Process for boreal and temperate forest countries.

Canada has been very supportive in all of these initiatives with a particular interest in the Montreal Process. This process was finalized last February when the ten participating countries met in Chile to write the Santiago Declaration, a document which includes the most comprehensive set of internationally agreed criteria and indicators to date. I had the honour of presenting this document at the Rome Forestry Summit.

Much of Canada's work to develop criteria and indicators, and a certification process, is driven by our commitment to sustainable forest management and its implications for international trade. With 10 percent of the planet's forests we account for 23 percent of world trade in forest products, worth some $45 billion a year.

CERTIFICATION

Through the Canadian Standards Association (CSA), we are developing certification standards that are compatible with the International Organization for Standardization. Internationally-accepted standards and the implementation of an objective certification process are essential for several reasons. They tell consumers that the forest products they purchase come from sustainably managed forests. They provide a universal yardstick against which suppliers can measure and demonstrate their forest management performance. And they help create a level playing field for global trade in forest products.

While achieving consensus presents a challenge, collectively, we are making progress. The opportunities to share technology, expertise and ideas are integral to developing a shared vision of sustainable forest management throughout the world. Work by you and your colleagues in the Model Forest Program is providing a global platform where these international initiatives can be tested and applied.

MODEL FORESTS

Canada is extremely proud of the Model Forest Network. It represents the kind of creative and innovative thinking that is going to take us into the next century in sustainable forest management.

Canada has 10 model forests, covering more than six million hectares of land. They represent six of the country's eight major forest regions. The purpose of the Model Forests is to promote partnerships and to encourage these partners to develop their own working vision of sustainable development.

In fact, the key to the success of the Model Forests is the multi-stakeholder approach. More than 300 organizations are involved in the Canadian Model Forests, representing governments, First Nations, industry, environmental groups, and other interested stakeholders.

Through our model forests, we are achieving success in a number of areas:

- The management of biodiversity and habitat conservation for endangered species.
- The stabilization of rural communities through the diversification of local economics and job creation.
- The development of new forestry practices for industry and small landowners.
- The innovative application of technologies such as Geographic Information Systems (GIS) and Decisions Support Systems (DSS).
- And we are also developing more effective public awareness and education programs.

I am particularly pleased with the progress across the Model Forest Network to address the role of aboriginal communities in sustainable forest management. I congratulate our hosts at the Prince Albert Model Forest, for their success in establishing a partnership that includes the First Nations, the Prince Albert National Park, the Province of Saskatchewan, the private sector and the NGO community. I am proud that the results they have achieved are based on a genuine desire and commitment to reach common goals. Indeed, the motto of the Prince Albert site-"working together, helping each other"-says it all.

A key goal of every model forest is to become self-sufficient–ensure, in effect, that the model created is viable and sustainable. To evaluate the best ways to achieve self-sufficiency, a complete and independent evaluation of the Canadian program will be conducted. I expect to receive the report by December of this year.

To date, more than 20 countries have expressed an interest in the Model Forest Network. As this Network grows and evolves, it has become a powerful tool to address global concerns, such as:

- increasing worldwide understanding of sustainable forest management,
- exchanging information and expertise, and
- testing and demonstrating innovative partnerships.

With each new country that comes on board, the Network dramatically increases in scope as it absorbs the different social, political, economic and cultural values of new members.

This growth reflects the increasing international awareness of the need for greater cooperation, participation and partnerships. That is why I am so pleased with the partnerships we have established with Mexico, Russia and Malaysia, and the interest of several other nations in joining the network.

I am also pleased to announce today that the United States is the latest link in the international chain. Three areas in the Pacific Northwest region of the U.S. have been added to the network. To those representatives from the United States who are here today, I extend a warm welcome.

I want to assure all of you that the Government of Canada continues to stand behind the Model Forest Network. We are committed to sharing expertise, information and other forms of technical support in the years ahead. We want to see a self-sustaining global network of model forests supported by participating countries that encompasses every major ecological region of the world. An expanded network will provide opportunities to share expertise through innovative approaches to sustainable forest management within a global framework.

CONCLUSION

The Government of Canada is proud of the momentum that we have helped to build toward sustainable forest management in this country and among other forest nations. But there is much more that we can do, and there is much more that we should do.

In two years' time, the world's Forest Ministers will meet again. I trust that at the second meeting of Forest Ministers, we will be even closer to international consensus on criteria and indicators, certification of forest products, and consensus on a definition of the term "sustainable forest management."

Your work through the Canadian and international model forest programs is making both a critical and practical contribution. Your efforts are proving that environmental, economic and social objectives can be integrated. This is the essence of sustainable forest management.

My challenge to you is to maintain your progress, and your dedication to cooperation and partnerships that are helping us learn from each other, and to share the benefits of new technology and other concepts that will make sustainable forestry a reality.

Future generations are counting on this progress. On behalf of the Government of Canada, I wish you every success during the rest of this conference, and in your respective model forest endeavours.

Thank you.

PART ONE:
GLOBAL CHALLENGES

Prince Albert
Model Forest
Association Inc.

MODEL FOREST
NETWORK
RÉSEAU DE
FORÊTS MODÈLES

Quo Vadimus?

Matt Heering

ABSTRACT. Deforestation has caused the collapse of past civilisations, including those of Easter Island, the Mayas of Central America, and the Sumerians of Mesopotamia. To avoid a similar fate, we should be prepared to undertake forceful and effective forestry action. This will entail food security and rural development; an increase in income for farmers and safe environments for indigenous peoples; empowerment of women and local groups; and family planning.

At an operational level forestry measures we should focus on include plantation forestry, agroforestry and the regeneration and rehabilitation of secondary forests. If such action is to succeed, and if global deforestation is to be avoided, it will require a well guided process of international cooperation. The internationalization of the Model Forest Program puts new impetus in this direction. *[Article copies available for a fee from The Haworth Document Delivery Service: 1-800-342-9678. E-mail address: getinfo@haworth.com]*

INTRODUCTION

I am grateful to the organizers for the invitation to deliver this address, putting me in a position to express my thoughts about international cooperation in forestry, a subject of utmost importance to the future of our world, in fact to our common future.

In doing so I will refer sometimes to TFAP, the Tropical Forestry

Matt Heering is Forestry Advisor to the International Affairs Division of the Ministry of Agriculture, Nature Management and Fisheries, The Netherlands.

[Haworth co-indexing entry note]: "Quo Vadimus?" Heering, Matt. Co-published simultaneously in *Journal of Sustainable Forestry* (Food Products Press, an imprint of The Haworth Press, Inc.) Vol. 4, No. 3/4, 1997, pp. 21-32; and: *Sustainable Forests: Global Challenges and Local Solutions* (ed: O. Thomas Bouman, and David G. Brand) Food Products Press, an imprint of The Haworth Press, Inc., 1997, pp. 21-32. Single or multiple copies of this article are available for a fee from The Haworth Document Delivery Service [1-800-342-9678, 9:00 a.m. - 5:00 p.m. (EST). E-mail address: getinfo@haworth.com].

Action Programme, the internationally coordinated framework launched in 1985 for collaborative action against unnecessary forest loss and degradation in the tropics. In TFAP, now adopted by over 90 countries, the donor community seeks to support national governments in their efforts to sustainably manage the forest resources and to develop its social, economic and environmental potential.

Over the past 10 years it has become clear that these objectives are relevant to all countries with forest resources, and that is why the "T" in TFAP has been changed into an "N," for National Forestry Action Plans. This is entirely in line with the call for action by United Nations Conference on Environment and Development (UNCED) in 1992, urging every government in the North and the South, in the tropics and in the temperate and boreal zones to develop a national plan for sustainable forest management.

National Forestry Action Plans have preceded the Model Forest Programs, but both initiatives have the same ultimate objective of sustainable development of forest resources.

I therefore will make an effort to demonstrate that intensive collaboration between the two programmes may have synergistic effects from which all participants will benefit.

Having said this, allow me to dwell briefly on 10 years of experience with the TFAP, an experience I share with two friends and colleagues present here, Salleh Nor of Malaysia and Ralph Roberts of Canada, both intimately involved in the design and implementation of the Action Plan.

Ten years after launching the Tropical Forestry Action Plan, it is very opportune indeed to check progress and discuss obstacles to effective cooperation of the international community in forest development. There are a few burning questions:

- Have we made an impact?
- Is forestry action contributing to poverty alleviation?
- Is sustainable development enhanced by trees and forests?
- Can economic growth be stimulated by forestry measures?

HISTORICAL CONTEXT

We all know that forests are indispensable for life on earth, but we know as well that deforestation and forest degradation have undermined the life support systems in many parts of the world, to the extent that living conditions have become extremely difficult. Nevertheless, the world community seems to ignore the very serious threats to the survival of our planet! Have we learned nothing from the disappearance of entire societies in the past? Mankind has destroyed civilizations before, also by ignoring the critical role of trees and forests for life.

Here are a few examples, taken from the book *A Green History of the World: Nature, Pollution & the Collapse of Societies*, by Clive Ponting. Easter Island is one of the most remote and inhabited places on earth. And yet the island had once a population of over 7000, with a flourishing and advanced society, with a tradition of monumental sculpture and stone work, known to us as *mohais* of which some 600 were made. Those monuments were used for burials, ancestor worship and to commemorate past clan chiefs. The statues were carved at the quarry, using only obsidian stone tools, and when finished brought to the place of destination; and that of course was an enormous undertaking. Each statue is some twenty feet in length and weighs several tens of tons.

There were no draught animals on the island and people had to rely on human power to drag the statues across the island. . . . using tree trunks as rollers.

Thus, Easter Island's solution to the transport problem proved to be the beginning of the end of their society. The small group of settlers which landed on the island in the fifth century had grown to 7000 people in the middle of the 16th century and the originally dense vegetation cover, including extensive woods, had been severely affected by the demand for tree trunks for sliding the heavy statues to the ceremonial sites across the island.

Massive environment degradation brought on by deforestation had a devastating impact on the social and cultural aspects of the society, bringing unrest and conflicts to a hitherto peaceful and flourishing community life. There were increasing conflicts over diminishing resources, resulting in a state of almost permanent warfare, slavery became common and as the amount of protein available fell, the population turned even to cannibalism. When the environment was ruined by self-inflicted deforestation the society very quickly collapsed with it, leading to a state of near barbarism.

About the Maya society a similar story can be told. The earliest settlements in Central America date from about 2500 B.C., with a slowly increasing population, which developed their astronomy and their highly complex but extremely accurate calendar. Huge pyramids, often aligned towards significant astronomical points, were built, but suddenly–after 800 B.C.–the society began to disintegrate. Slash and burn practices and shifting cultivation had been the highly stable agricultural system which provided sufficient food and fuelwood for relatively small populations. But when the population gradually grew to a peak of 5 million, and with the fallow requirements of some 20 years, the system could not be productive enough to support such a large population. Consequently the jungle at the hillsides was cleared and fields created, with terracing to try to contain the inevitable soil erosion. However, the ecological basis to support a massive agricultural production system was simply not there because the fertile volcanic soil is highly vulnerable to erosion once it is cleared of trees.

This situation was aggravated by the ever increasing production of huge amounts of lime plaster that decorated the many ceremonial buildings, for which colossal quantities of fuelwood are consumed. And thus, the collapse of the Maya civilization was also caused by a self-imposed ecological disaster, triggered off by large-scale deforestation to provide the Mayas with land for food production and fuel for lime stucco production.

Another example is that of the Sumerian society, which developed about 3000 B.C. in the Mesopotamian desert. Why was it that the population of this first literate society in the world dwindled to nothing?

The detailed administrative records kept by the temples of the city states provide a record of the changes in the agricultural system and an insight into the development of the major problems that led to the downfall of this once vast granary and empire.

And the short answer to the question is that the Sumerians themselves destroyed the world they had created so painstakingly out of the difficult environment of southern Mesopotamia, at the one hand, by clearing the richly forested country to provide fields for agriculture, at the other by introducing irrigation.

The explanation is that the twin rivers, the Tigris and the Euphrates, were at their highest in the spring following the melting of the winter snows near their sources and at their lowest between August and October, the time when the newly planted crops needed the most water. Water storage and irrigation therefore were essential if crops were to be grown.

But with summer temperatures often up to 40 degrees Celsius, evaporation from the surface increased considerably and with it the amount of salt in the soil. Wheat and barley were the main crops, but the harvests gradually declined following environmental degradation caused by erosion due to deforestation and by salinization as a consequence of irrigation. Modern agricultural techniques dictate to leave the land fallow, but rising populations and the limited availability of land that could be irrigated made this impossible and brought about disaster. The artificial agricultural system that was the foundation of Sumerian civilization proved to be unsustainable and in the end, irrigation in combination with the destruction of the woods and forests around the settled communities, brought about its downfall.

MODERN CHALLENGE

It is not difficult to find more examples of self-inflicted environmental collapses, caused by the pressure from growing populations, and although humans in the past have succeeded in obtaining more food and extracting more resources for increasingly complex and technologically advanced societies, we must pose the following question: Have we been any more successful than the Easter Islanders, the Mayas or the Mesopotamians in finding a way of life that does not fatally deplete the resources that are available to us and that irreversibly damage our life support system?

In accepting the challenge of sustainable development, our personal involvement in and commitment to international forestry cooperation, four verbs cross frequently our minds: hope, undertake, persevere and succeed. Our Prince of Orange, who lived from 1533 till 1584, and who was, so to say, the founder of our present royal family and of our democratic society, apparently has faced similar challenges as we do. He spoke the famous words (in French): *"Point n'est besoin d'esperer pour entreprendre, ni de reussir pour perseverer"* or, in English, "One need not hope in order to undertake, nor succeed to persevere." There is no better way of describing our global struggle for developing and conserving forests to the benefit of mankind. Will we succeed? At the one hand there is enthusiasm, at the other frustration. On the one hand there is cooperation, on the other, selfishness. On the one hand we see some real progress, on the other, setbacks due to political considerations and decisions which go often beyond our understanding.

In spite of the international efforts to contain deforestation, the outlook continues to be grim; we are about to destroy our very own civilization! Again questions: what errors did we make? Has the collapse of the global environment already set in, with global warming and the irreversible destruction of natural resources and biodiversity the frightening signals of coming disaster?

This leads to the very personal question: Should we be witnessing disasters or make an unprecedented effort to convince the world that it is high time to undertake forceful and effective forestry action?

The answer of course is that we all have the personal responsibility to act rather than to watch. What we need is substantive action rather than political debate. What we need is food security and rural development. What we need is income increase for farmers and safe environments for indigenous people. What we need is empowerment of women, and of local groups which seek to escape lifetime poverty and misery. What we need, yes let me say that loudly and clearly, is family planning. And above all, what we need is strong political commitment, dedicated donor assistance and perseverance!

Shaping the future: what can we do?

Do we have our priorities right? Have we learned our lessons? Let us take a hard look at the facts:

The world population will double in the next 25 years and so will the demand for food, fuelwood and all other forest products.

In the past 10 years we have achieved a lot and UNCED's call for sustainable development and forest management has invigorated our collective efforts towards more effective international cooperation in forestry, with your Model Forest Programme as a striking example of much needed collaboration among partners of different backgrounds. National Forestry Action Plans (NFAPs), developed under TFAP in over 90 countries is another encouraging indication that we are on the right track. But we have not been able to significantly reduce deforestation, let alone to increase the world forest cover.

How are we going to cope with the extremely difficult and complex task to let forestry be the very successful contributor to effective improvement of living conditions?

The first thing I would say we have to do is to reconsider our priorities, learning from the failures and errors of the past.

Three priority areas stand out:

1. We have to concentrate on the people concerned (the rural poor and the indigenous people living in and around the forests), because they need to do the work, they must produce food, fuelwood and other forest products, they decide on their priorities and on them depends effective action.
2. We have to enhance better fieldwork: effective, successful and sustainable.
3. We have to promote and support local leadership, namely that of local chiefs and coordinators, of farmers, foresters and agroforesters.

You see, the theme of the conference is well taken. After many years of trying, we seem to realize that in order to successfully address global problems we need to find local solutions.

The next question is: "On which forestry measures we should focus if we want to make a sizable impact?" Again: I would aim at three priority areas.

- *Plantation forestry*: Since we are still losing forest at a rate of some 15 million hectares per year, we need to do what we internationally have agreed to do, to plant trees and forests in order to reach a net increase of forest cover of 12 million hectares per year (this is one of the objectives of the climate convention).

This translates into an annual re-afforestation of at least 27 million hectares. That seems a colossal task, but if each country would do its share it is definitely within reach.

- *Agroforestry*: There is a huge need to enhance the planting on farms and homesteads of trees in combination with food crops, providing the rural people with technologies to increase their income and decrease soil erosion.

ICRAF trials have proven to be extremely promising, and research for development seems to pay off well, for example, in Peru where shifting cultivators in the Amazon area are provided with alternatives for slash and burn practices and can earn up to ten thousand dollars per hectare per year, by horizontal and vertical diversification and optimal use of the spatial area.

- *Regeneration and rehabilitation of secondary forests*: There is an enormous potential in secondary forests, which in many countries only to a very limited extent is developed. If we want to take the pressure off the virgin forests, the first thing we have to do is take concrete action to regenerate and rehabilitate the logged-over forests, which is of course the normal procedure in Canada, and one of the targets of the Model Forest Program.

Obviously, the above mentioned areas for priority action are not self-standing, nor can any action be undertaken without multi-disciplinary discussion. But let us avoid the endless and fruitless debates on multi-disciplinarity of the past. A practical approach always takes into consideration the influences of developments outside the forestry sector. Rather we must concentrate our collective efforts on support to developing sustainable forest management, assisting in defining realistic national targets, in preparing a timetable for increasing the forest cover and in improving forest quality by improving forestry practices.

Critically important is the volume of financial resources available to the forestry sector, as is the need for policy reform and legislation to obtain better revenues from the forests.

And last but not least there is the urgent need to recommit the international community to the development and conservation of forests. How to orchestrate this is difficult to say, but a vision must be developed and to that extent I would like to submit a few ideas which may help in choosing the right direction for the next decade.

I will put it bluntly. Without a well developed and well guided process of international cooperation there is no hope for effectively addressing and resolving the problem of global deforestation; which means that unless we drastically step up our collective efforts, the gradual but definitive destruction of our society, of our planet, would become unavoidable.

If we don't want this to happen, if we don't want to become victims of our own short-sightedness, if we don't want to become the Easter-Islanders of our time we have to get our act together, in particular in international cooperation.

We need urgently to enhance dialogue, partnerships and collaboration. We need urgently to start respecting national and local priorities. We need urgently to provide unselfish assistance to the national efforts towards sustainable development.

Such an undertaking, of recommitting ourselves to international forestry cooperation, requires courage, vision and leadership.

Unfortunately, during the past 20 years, those characteristics (courage, vision and leadership) have not been displayed by the international organizations which were created to lead the international community towards effectively meeting global challenges. This already scandalous situation worsened with the recent cuts in many if not all aid budgets, often limiting international cooperation to bilateral efforts.

There is an urgent need to reverse this trend!

What is it that we have to do?

First of all, we have to organize ourselves, taking the recent decisions of the CSD as guidelines, not waiting any longer for leadership, but developing leadership while working.

Of utmost importance thereby is that the international community stand ready to assist and support the national and local authorities responsible for the forestry sector.

I am very much encouraged by the internationalization of the Canadian Model Forest Program. What you have developed in the various model forests in Canada in regard to common approaches and understanding is of great importance to present and future partners in the search for sustainable management of forests.

By launching the International Model Forest Program, with a

special secretariat to coordinate and support the efforts, Canada puts new impetus into the existing framework for collaborative action.

It is clear that the sustainable management of not only tropical but as well of temperate and boreal forests has become the focus of worldwide forestry action. What is now called National Forestry Action Plans (NFAPs) applies to all countries and it builds on ten years of experience, in which every individual country is entirely free to define and implement its priorities towards forest development. This is also the approach taken in the Model Forest Program, whereby, in dialogue with the local people and based on their best professional judgement, criteria for sustainable management are defined and developed.

In this respect I see the NFAP as a viable, troublefree and collaborative framework, within which the International Model Forest Program can thrive, prosper and be delivered effectively and efficiently. And we couldn't serve the international community better than by creating converging mechanisms, in order to take maximum benefit from the lessons learnt in TFAP and in the Canadian Model Forest Program.

Some of those lessons surfaced recently in The Hague, during a workshop organized by CIDA and the Dutch Ministry of Agriculture, Nature Management and Fisheries. The auto-evaluation by some 30 National Coordinators of the NFAP-process led to some very direct and practical recommendations to the United Nations Commission on Sustainable Development, which recently met in New York to discuss the global forestry situation.

Coming from the field level and from the people who are deeply involved in the day to day management of forests, these recommendations represent in fact clear instructions to the partners in the NFAP-process and in the Model Forest Program. Shouldn't we act accordingly? The international community cannot ignore these recommendations!

National Coordinators have given us in a nutshell what we have to do in order to revitalize international cooperation in forestry. Not a difficult task if the good will and the political commitment are there; and the conviction that this is the moment to take urgent action in order to maintain the global perspective for resolving a gigantic global problem, that of environmental destruction and potential collapse.

There is that old saying that it is cheaper and easier to prevent than to repair, and that applies undoubtedly to our environment.

Timely forestry measures, plantation forests, agroforestry and regeneration and rehabilitation of secondary forests will definitely help to increase rural incomes, improve living conditions, save forests from wasteful exploitation and conserve biodiversity.

What are we waiting for? Couldn't Canada and the Netherlands (and possibly other countries) jointly take the initiative to set up a support centre for global forestry action; for generating renewed commitment to the development and sustainable management of forests; for mobilizing the international community and supporting the local and national efforts; and for servicing the CSD and its Forest Panel in the difficult task to harmonize the global endeavours towards sustainable forest development?

What are we waiting for?

Mr. Chairman, we can not afford to wait and see what the political levels decide to do, we have already spent too much time on high-level international congresses and expert meetings. It is urgently necessary to dictate our own marching orders.

As the Prince of Orange said, "One need not hope in order to undertake, nor succeed to persevere!" And he defeated the Spaniards after a war that dragged on for 80 years.

So, in summary, here are my suggestions for the 10 commandments for the next decade of international forestry cooperation:

1. Aim at direct assistance to the rural and indigenous people concerned, and to the NFAP and Model Forest Program coordinators.
2. Aim at concrete results at the field level, by enhancing effectiveness and sustainability of action.
3. Promote and support local leadership.
4. Undertake plantation forestry and regeneration of logged-over forests as the most efficient way to produce timber.
5. Enhance agroforestry action in order to mobilize the rural population for effectively increasing their income, reducing soil erosion and decrease forest destruction caused by slash and burn practices of shifting cultivators.

6. Promote the rehabilitation of secondary forests for the production of forest products.
7. Recognize, strengthen and support the national and local approaches as the genuine efforts towards sustainable forest development.
8. Cooperate at the national, sub-regional and international levels in order that through coordinated action best possible use can be made of the available resources.
9. Reallocate financial resources so to favour forestry action and to close the gap between the project plans and available funding.
10. Increase political commitment in order to successfully initiate or strengthen policy reforms for dynamic forest development.

Is that all, Mr. Chairman?

No, there is one more burning question–that of family planning, an apparently sensitive issue.

But if we continue to deny the seriousness of this issue, we will indeed have doubled the world population in 25 years.

We had better double our efforts to avoid that, because it will seriously jeopardize our collective efforts to bring deforestation and environmental degradation under control. And also here the theme of the conference applies: global challenges, local–and even family–solutions.

To conclude, Mr. Chairman, ladies and gentlemen, I would like to recall the words Churchill spoke in the beginning of the second world war, words that come close to those spoken by the Prince of Orange, centuries ago. Churchill gave the following definition of success: "Success is going from failure to failure without losing enthusiasm."

I would like to paraphrase this into a more optimistic description of success. Success for us is creating so many convincing model forests, that politicians around the world bring back forests to the top of their agendas, and commitment and money to global forestry action.

Thank you for your attention.

REFERENCE

Ponting, C. 1992. *A Green History of the World: Nature, Pollution & the Collapse of Societies*. St. Martin.

Sustainability:
The Panacea for Our Forestry Ills?

M. N. Salleh

ABSTRACT. Fewer than one tenth of tropical forests are being managed on a sustainable basis. Sustainable forest management means managing the forest in such a way as to not irreversibly reduce the potential of that forest to produce all products in subsequent harvests. The United Nations Conference on Environment and Development in Rio resulted in several decisions that are relevant to the future of forestry. The Conference also focused world attention on questions of the environment. One outcome of this increased awareness has been the growing support for eco-labelling, which may provide an opportunity for those countries able to prove their forest products are harvested sustainably. Other economic opportunities present themselves in the utilization for cellulose of tree crops such as rubberwood and oil palm trunks and fronds. Non-wood resources such as rattan also hold promise if we are able to grow them in conjunction with existing tree crops. The roles of tropical forests as carbon sinks require more in-depth study as does the question of what constitutes critical levels of biodiversity. Aesthetic values such as recreational use increasingly require that sufficient buffer zones of unique features be preserved. These challenges demand that the forestry profession becomes more proactive and support major policy changes to address the need for sustainable forest management. *[Article copies available for a fee from The Haworth Document Delivery Service: 1-800-342-9678. E-mail address: getinfo@haworth.com]*

M. N. Salleh is President of the International Union of Forest Research Organizations and Director General of the Forest Research Institute of Malaysia.

[Haworth co-indexing entry note]: "Sustainability: The Panacea for Our Forestry Ills?" Salleh, M. N. Co-published simultaneously in *Journal of Sustainable Forestry* (Food Products Press, an imprint of The Haworth Press, Inc.) Vol. 4, No. 3/4, 1997, pp. 33-43; and: *Sustainable Forests: Global Challenges and Local Solutions* (ed: O. Thomas Bouman, and David G. Brand) Food Products Press, an imprint of The Haworth Press, Inc., 1997, pp. 33-43. Single or multiple copies of this article are available for a fee from The Haworth Document Delivery Service [1-800-342-9678, 9:00 a.m. - 5:00 p.m. (EST). E-mail address: getinfo@haworth.com].

33

THE CONCEPT

Sustainability is the catchword which is widely used nowadays. The term "sustainable forest management" is a integral part of the language of forestry and has been in place for a long time, much earlier than before the words "sustainability" and "sustainable development" became household words as a result of the Report of the World Commission On Environment and Development in 1987.

Forestry is implicitly a long-term venture and since forestry involves managing a biological renewable resource, the concept of sustainable forest management has been, from the beginning of forestry, the basic foundation, pillar and backbone of the principle and the science of forestry and a principal measure of good forestry practices.

It is therefore an unfortunate state of affairs and a poor reflection on the practice of forestry when questions are asked as to the "sustainability" of forest management. The "sustainability of forestry" practices are being questioned, especially of tropical forests. A study by the International Tropical Timber Organization (ITTO) (Poore et al. 1989) reported that less than one-tenth of tropical forests are managed on a sustainable basis.

This is a blow to forest management, and professional foresters in tropical countries should embark on a soul-searching examination to determine why the basic pillar of forestry has crumbled and is being questioned.

There are many reasons and causes for such a situation, but more importantly, immediate steps must be taken to remedy the situation. Wyatt-Smith (1987) and Poore et al. (1989) discuss the issue in some detail.

"Sustainable forest management" can be defined as "the management of forests that will not irreversibly reduce the potential of that forest to produce all products in subsequent future harvests." In layman terms, it is to utilize the forest while ensuring that the productive capacity of the forest or the ability of the forest to regenerate itself is not impaired and that the forest can continue to produce the desired products at a determined and usually enhanced level, in perpetuity.

In practice, in the management of tropical forests, the concept of sustainability operates at several levels. Thus, there is production of goods, services and benefits and maintaining environmental and conservation benefits, yields of timber and non-timber products and the generation of income over generations, without the degradation of the environment or the productive capacity of the forests.

With the concern of biological diversity and the role of forests in the protection of the climate, the concept of sustainable forest management has to be expanded from the traditional production of timber only to the conservation and the sustainability of the biological wealth of the forests as well as to the environmental role of forests. This becomes a much more complex issue.

There are various critically important issues in the discussions of sustainability that must be considered.

- *Level:* What is the level of quantum of production that must be sustained? Should production be at current levels or a level determined by the productive capacity of the forest?
- *Time:* What time frame is being considered to be the sustainable rotation or cycle? A rotation set at 30, 60 or 100 years will have different influences on sustainability.
- *Product:* What product is being produced on a sustainable basis? Production of sawn logs or plywood logs or pulpwood will affect sustainability. Quality of product is thus an important consideration.
- *Non-Wood Products:* Should non-wood products such as rattan, bamboo, water and others be considered? How can they be integrated into the concept of sustainable management of forests?
- *Biodiversity:* How can biodiversity, including fauna and flora, be managed on a sustainable basis? This is of increasing importance, especially for tropical forests.
- *Environment:* Can the environment of forests, including such intangible values as carbon sequestration, aesthetics and recreational use, be managed on a sustainable basis?
- *Space:* The extent of the coverage of concern, whether at a local, regional or national level.

There are many definitions of sustainable forest management, many of which are similar. Mok and Poore (1991) defined sustainable forest management as "the process of managing forest land to achieve one or more clearly specified objectives of management without undue reduction of its inherent values and future productivity or undesirable effects on the physical and social environment."

FORESTRY AND UNCED

Of relevance to forestry are decisions arrived at the United Nations Conference on Environment and Development (UNCED) at Rio de Janeiro in June 1992 on the following:

- *Agenda 21:* a statement of goals and objectives, as well as a list of strategies and actions, that are to be taken to implement the principles enunciated in the Earth Charter (a document of principles of the Conference);
- *Convention on Biological Diversity:* developed and finalized for signing at Rio;
- *Framework Convention on Climate Change:* developed and finalized for signing at Rio;
- *Statement of Principles on Forest:* a nonbinding agreement for the protection of all types of forest.

A whole chapter in *Agenda 21* addresses the problem of combating deforestation. The issues of sustaining the multiple roles and functions of forests, of greening degraded areas through forest rehabilitation, afforestation, reforestation and other rehabilitative means, and of increasing the value of forests by increasing the production of goods and services, in particular, the yield of wood and non-wood forest products, and by ecotourism and the managed supply of genetic materials, are given particular attention in this chapter.

Agenda 21 also devotes one whole chapter to the conservation of biological diversity although this subject is dealt with in greater depth in the *Convention on Biological Diversity.* Biodiversity, on the global scale, is declining due mainly to human activities such as the destruction of habitat and by overharvesting of some ecosystems. The chapter in *Agenda 21* deals with strategies to protect biological diversity and to use biological resources sustainably.

The *Convention on Biological Diversity*, as does *Agenda 21*, places great emphasis on the conservation of biological diversity and the sustainable use of its components. As tropical forests contain more than half of the world's biological diversity, with some estimates being as high as 80-90 percent, the responsibility to ensure protection of tropical forests is high indeed. Issues of developing national strategies, plans or programmes for biodiversity, of establishing a system of protected areas for biodiversity, of maintaining viable populations of species in natural surroundings, of rehabilitations and restoring ecosystems, of legislating provisions for the protection of threatened species and populations, of managing relevant processes and activities that adversely affect biodiversity, and of *ex situ* conservation of biodiversity, are those relating to forestry that have been brought into focus by the convention.

The *Framework Convention on Climate Change* makes reference to forests and forestry in several contexts. The need to control, reduce or prevent anthropogenic emissions of greenhouse gases in the forestry sector is stressed. The need to conserve and enhance forests as a sink and reservoir of greenhouse gases is referred to.

No legal instrument on forestry was developed for signing at UNCED due to widely differing views between developed and developing countries. Nevertheless, a *Statement of Principles on Forests* was agreed upon at Rio, and this may well be a precursor to a legal instrument being developed in the future.

The *Statement of Principles on Forest* is clear in the multiple and complementary functions and uses of forests, and stresses that forest resources and forest lands should be sustainably managed to meet the social, economic, ecological, cultural and spiritual human needs of present and future generations. The principles in the Statement apply to all types of forests, boreal, temperate and tropical.

The *Convention on Desertification* has recently been signed, over two years after Rio. Forests and trees also play an important role in this convention.

CHALLENGES OF SUSTAINABLE MANAGEMENT

Rio placed the issues of the environment firmly on the global map. The gathering of 104 Heads of State ensured that world attention was focused on Rio. The mass media reached into homes in all corners of the earth in vividly reporting all activities at the UNCED.

The issue of forests figured prominently throughout the period of the UNCED because of the significant role they play in maintaining climate, in being major storehouses of biodiversity, as well as in maintaining structural and functional biodiversity at the ecosystem level. It would be justifiable to say that public awareness of the importance of forests was increased as a result of Rio. Because the mass media is more efficient in reaching the layman in developing countries, the main consumers of our tropical timbers would be increasingly amenable to calls that timber and timber products from forests that are not managed sustainably be boycotted.

An anti-tropical hardwood timber campaign has already been waged for several years now by powerful non-governmental organizations (NGOs) and lobby groups in developed countries. These bodies have been able to exert influence on certain governments and consumers alike. An evolution of this campaign is eco-labelling of timber products and in the wake of Rio, this phenomenon is likely to gain prominence and eventually become part and parcel of sustainable forest management and development.

Eco-labelling could in fact work in a country's favour if it can be proven that forest management practices in that country are indeed sustainable. Then her forest products would be widely accepted by a world market that is increasingly environment conscious, maybe even at a premium.

OTHER CELLULOSIC RESOURCES

In many tropical countries, there are other tree crop plantations which can be sources of wood. In Malaysia, rubberwood has developed into a major wood resource over the last decade and will continue to do so. In this respect, the rubber planation sector has recognized its contribution and responsibility to the rubberwood industry. There is a definite psychological and policy change in the rubber plantation sector to accept that rubberwood is not a "by-product" of their industry but a valid and economic product which has to be incorporated into the management system of the rubber plantation.

The oil palm industry is also potentially an important source of cellulosic material which needs to be encouraged to embark on manufacturing industries based on oil palm trunks, fronds and empty fruit bunches.

By all definitions, rubber and oil palm plantations are "forests" and supply of cellulosic materials from these resources should be considered in the sustainable supply equation for the country.

NON-WOOD RESOURCES

Forest management has too long been "timber-oriented" and has not considered other non-wood resources which are becoming increasingly important and which have tremendous potential in economic contribution. While it is impractical to cater for the requirements of all non-wood resources, it is possible and potentially economical to consider some of the more important resources from the forests. For example, it is possible for forest management to integrate growing of rattan within the management of trees.

Since rattan can reach economic maturity at the age of 15 years, it is suggested that two rotations of rattan can be integrated into one rotation of timber under the SMS system or three or four rotations of rattan into the MUS system.

This system involves the intensive planting of rattan along rows as well as the management of naturally growing rattan. While the economics of this system need to be examined, there are adequate preliminary indications that planting rattan is economically viable, as long as the land is free. However, how can rattan and other non-wood resources, such as bamboo, resins, medicinal plants, and others, be included into the sustainability equation?

TROPICAL FORESTS AS CARBON SINK

The role of forests as a carbon sink is a very complex issue and the purpose of raising this issue is to sensitize managers and policy-makers on the increasing importance of this issue. The carbon sink role of tropical forests is well established and the need to sustain this capacity must be considered.

Unlike a monoculture plantation, the complexities of species, tree sizes, plant communities and vertical and horizontal structural differences make the understanding of the complex carbon sink role difficult, especially understanding the changes resulting from the logging of such forests. Nevertheless, the significant contribution of tropical forests in the global environment balance must be kept in mind as we harvest our forests but a more in-depth study is required in order to understand the complex phenomena that exist in the forest ecosystems. Current debate on "Joint Implementation" requires an in-depth study of the role of forests in carbon sequestration.

BIOLOGICAL DIVERSITY

The issue of biodiversity is becoming global where the rich biodiversity of life in the tropical forest is even being considered a common heritage of man. What are the impacts of logging on biodiversity? Will there be loss of biodiversity, and if so, how crucial will this be? What is the critical point beyond which further removal of timber in a logging operation leads to breakdown in structural and function biodiversity that leads to forest degradation and eventually to deforestation?

The removal of high-value economic species could result in a long-term degradation of the genetic diversity. How can this be prevented? One of the most urgent aspects with regard to biodiversity is to document and understand the actual biodiversity that exists and its role in maintaining the ecosystem.

This is an enormous task and if we are to seriously address biodiversity, an evaluation of what we have and where, is most urgent. How can we truly harness the potential wealth that exists without our biological diversity? Tropical countries have to develop the technical capability to do so and not be solely dependent upon the developed world.

AESTHETIC VALUE OF FORESTS

Demand for the recreational use of forests will increase and it is therefore critical that sites of unique recreational features, such as waterfalls, caves, "big tree plots" and areas of unique interest, be protected now. Protection of such upstream facilities is critical for long-term sustainability. Lest these facilities become isolates in a sea of development areas that eventually could erode the existence of these very facilities, a buffer zone of forested area of sufficient size will need to be retained.

NGOs AND FOREST MANAGEMENT

With the global concern on environment and after UNCED, forestry has found itself in the global limelight, especially tropical forestry. Non-governmental organizations (NGOs), both national and international, have voiced concern on forestry practices globally, especially in tropical forests. Besides the issue of sustainability and biodiversity, the issue of forest dwellers has been highlighted.

International NGOs exert great influence on the policies of important countries and some countries and agencies in the West have banned the use of tropical timber. The call for sustainable management of tropical forests is loud and clear and the international movements will have a far reaching impact on forestry practices in developing countries.

ROLE OF THE FORESTRY PROFESSION

A conference of senior foresters in Yokohama in 1991, organized by ITTO and the Forestry Agency Japan, emphasized the role of the forestry profession in sustainable forest management. This is a critically important aspect in the overall sustainable forest management scenario.

Without true professionals, the practice of the profession will not be sustained. There needs to be greater professionalism within the forestry profession. The international and global interest on tropical forestry is both a challenge and an opportunity.

Some view it as a threat. The opportunity for forestry is now. Governments and policy-makers are more aware of forestry, its potential and its role. Government and policy-makers are also aware of their own responsibility. This awareness needs to be translated into a commitment.

This opportunity is not only global. The profession needs to be proactive and innovative. "Business as usual" will not be adequate. Major policy changes are necessary. The future of forestry is intrinsically linked with agriculture, land use and the industrial sector. These opportunities open new challenges.

Come 2000, there will be timber for the consumption of our nation, but of what type and at what price? There are so many issues that need to be examined and solutions depend much on the foundations that we set now. Difficult decisions need to be taken now to ensure sustainability of forestry in the year 2000.

CONCLUSION

Whatever else may be the outcome of decisions made at Rio, the topic of forestry, much debated in the run-up to and at Rio, is likely to become a burning issue in the years to come. Public opinion is likely to grow stronger for the protection of forests worldwide, whether they be boreal, temperate or tropical. Consumers in the more affluent societies will increasingly shun products that come, or are believed to come, from forests that are not sustainably managed. These consumers are likely to pay more for products that come from sustainably managed forests.

It is the moral duty of nations to ensure that all our natural resources, including forests, are sustainably managed so that the needs of our future generations are not compromised. Insofar as forests are concerned, management plans in place will have to be strictly enforced, and the political commitment needed to do this will have to be duly provided. The nation's efforts in this direction should not be because of Rio or of international pressures. Rather, Rio should just be a reminder that forests are a resource that needs to be protected and sustainably utilized.

Many tropical countries are still fortunate to be amply endowed with such a valuable resource as the rainforest which has taken about 130 million years to evolve to the present complex structure. It is a natural resource that is really being held in trust for future generations by the present generation. Utilization of this resource should be sustainable such that the impact on the environment is minimal and the quality of life not threatened. This will ensure that we fulfil our obligations to future generations.

UNCED was the forum where our leaders expounded the responsibilities of all nations with regard to the global environment and forests. International eyes are watching. With ITTO setting "Target 2000" at which only timbers from "sustainably managed forests" can be traded on the international market, there is now a rush to develop "criteria and indicators" for sustainability. Forestry is now faced with an adversity it has never faced before. As someone said, adversity is the "springboard to great achievement." Indeed, the pressure to achieve sustainability may yet be the panacea to all forestry ills of the past, but are we too late?

REFERENCES

Mok, S. T. and Poore, D. 1991. Criteria for Sustainable Forest Management. Report prepared for ITTO.

Poore, D. et al. 1989. No Timber Without Trees. Earthscan Publications Ltd., London.

World Commission on Environment and Development. 1987. Our Common Future. Oxford University Press.

Wyatt-Smith, J. 1987. The Management of Tropical Moist Forest for the Sustained Production of Timber: some issues IUCN/IIED Tropical Forest Policy Paper No. 4, IUCN.

Cultural Values versus Science

Chris Maser

ABSTRACT. The history of rationalistic thought in the west has been built on paradigms that provide justification for the fragmentation and domination of nature. We are therefore faced with a crisis in perception that overshadows any discussion of environmental issues. Conservatives tend to be more resistant to change than liberals. As a result, they are more likely to place primacy on others of one's race, creed and religion, as well as one's personal needs. Conversely, it is less likely that they will have regard for the land's sustainable capacity. Liberals tend to take a systems approach to thinking and are therefore less inclined to tinker with the individual components of a system. Science alone cannot reconcile opposing points of view which are based on value, because science was not designed to deal with values. Conducting a credible search for the truth demands that we first recognize the limitations of our own thinking, both in our tendency to form a single hypothesis and our tendency to be "method oriented" rather than "problem oriented." Although science can never accurately represent nature, it has, over time, piqued our imagination, challenged old ways of thinking and demonstrated the fuzziness of our world view. *[Article copies available for a fee from The Haworth Document Delivery Service: 1-800-342-9678. E-mail address: getinfo@haworth.com]*

THE REDUCTIONISTIC MECHANICAL WORLD VIEW

To understand the present conflict between our relatively sedentary cultural values and the ever-changing views of science, we

Chris Maser is Author, Lecturer and International Consultant on sustainable forestry, Oregon, USA.

[Haworth co-indexing entry note]: "Cultural Values versus Science." Maser, Chris. Co-published simultaneously in *Journal of Sustainable Forestry* (Food Products Press, an imprint of The Haworth Press, Inc.) Vol. 4, No. 3/4, 1997, pp. 45-52; and: *Sustainable Forests: Global Challenges and Local Solutions* (ed: O. Thomas Bouman, and David G. Brand) Food Products Press, an imprint of The Haworth Press, Inc., 1997, pp. 45-52. Single or multiple copies of this article are available for a fee from The Haworth Document Delivery Service [1-800-342-9678, 9:00 a.m. - 5:00 p.m. (EST). E-mail address: getinfo@haworth.com].

45

must consider the role of those rationalistic thinkers who gave frame and body to a reductionistic mechanical world view over a period of some 300 years.

Rationalistic thinkers, such as Francis Bacon (1561-1626, English philosopher and essayist), Galileo Galilei (1564-1642, Italian scientist and philosopher), Rene Descartes (1596-1650, French philosopher and mathematician), John Locke (1632-1704, English philosopher), Isaac Newton (1642-1727, English mathematician, scientist, and philosopher), and Adam Smith (1723-1790, Scottish political economist and philosopher), legitimized and institutionalized the lust for material wealth over which feudal society had for so long fought. In so doing was born the reductionistic mechanical world view.

Consider the collective paradigm of these renowned men: "Nature's sole value is in service to the material desires of humanity" (Bacon, Locke). "But Nature must be tortured before Her secrets will be revealed for human use" (Bacon). "Once wrested from Nature, only those secrets that are measurable and quantifiable are real or relevant and can be studied" (Galilei). "Because real things are both measurable and quantifiable, they must operate through predictable linear mechanical principles, like an enormous machine" (Newton). "And like a machine, real things can be understood by disassembling the things themselves into smaller and smaller, more manageable pieces, which can then be rearranged in an order deemed logical to the human mind" (Descartes).

With reductionistic mechanical logic, major segments of Western society confer upon themselves the unlimited rights of individual private property (Locke) for which people must compete with one another in pursuit of their own self-interests (Smith). Such self-interest is to be free from any government interference because the "Invisible Hand" of moral guidance will temper self-interest in the pursuit of material wealth–for the betterment of society (Smith). While Smith's "Invisible Hand" has spiritual connotations, they are out of keeping with his pursuit of self-interests in the form of material wealth. Further, his notion of a "Higher Moral" automatically guiding human action was already overshadowed by the accepted reductionistic mechanical posits of Bacon, Galilei, Descartes, Locke, and Newton.

The current analytical perspective thus involves a four-part process: (1) disarticulate the system into its component parts, the smaller the better, (2) study each part in the sterility of isolation, (3) glean a knowledge of the whole by studying the parts, and (4) rearrange the parts in a way that logically fits our reductionistic mechanical world view. This is like disarticulating a live cat, rearranging its parts, putting them back together again, and expecting the cat to live and function as before.

Two implicit assumptions are based on this analytical perspective: (1) systems exist as aggregates of interchangeable parts functioning in a predictable linear fashion, and (2) by optimizing each part to a desired human end, the entire system is optimized for human benefit. We thus continually fragment our problems into smaller, more manageable (albeit linear and increasingly dysfunctional) pieces while our challenges become ever-more systemic.

A CRISIS IN PERCEPTION

Our behaviour, the extension of our feelings, thoughts, and values is destroying our biosphere because we insist on applying the reductionistic mechanical model as it was falsely legitimized many decades ago. But it is increasingly apparent that the root of our social/environmental crisis is of spiritual/moral values brought about by clinging to a self-centred world view and its associated value system.

Under the influence of the reductionistic mechanical world view, it is too easy to dismiss as impractical idealism any attempt to refocus from immediate bread-and-butter issues to long-term processes and futuristic ideas. Further compounding the belief that long-term processes and futuristic ideas are merely impractical idealism is the notion of "conversion potential." For many people, the only value of anything is its "conversion potential," which dignifies with a name the concept that nature, having no intrinsic value, must be converted into money before any value can be assigned to it. All nature is thus seen only in terms of its conversion potential.

Western civilization was lulled to sleep by the thinking of Descartes and Newton, but we are awakening to the flawed nature of their premises. The new vision of a single organic whole is being derived through the revolution in physics, primarily quantum mechanics and the work of Albert Einstein et al. But conservative thinking has yet to catch up with the knowledge of modern physics and a changing world view.

No matter what central issue is discussed, therefore, the dynamics are the same–an underlying crisis of perception. Our continued acceptance of this world view as the absolute truth and the only valid way to knowledge has led to the current global crisis and propels us ever closer to social collapse through environmental destruction.

Now, as history's veil enshrouds the events of the twentieth century, the cherished cultural values of the reductionistic mechanical world view are in deadly grapple with the revelations of science that increasingly challenge that view. One of the major problems facing us today is the way in which we accept that challenge, be it from a conservative frame of reference or that of a liberal, and all the shades in between. The differing perspectives define the terms of debate.

CONSERVATIVES AND LIBERALS: A DIFFERENCE OF VALUES

In my experience as a facilitator helping to resolve environmental conflicts, the more conservative a person is the more resistant he or she is to change, seeing it as a condition to be avoided because he or she feels a greater sense of security in the known elements of the status quo, especially where money and private property are concerned. But, as Helen Keller once said: "Security is mostly a superstition. It does not exist in Nature. Life is either a daring adventure or nothing." Conversely, the more liberal a person is the more likely he or she is to agree with Helen Keller and risk change on the strength of its unseen possibilities.

A conservative person is likely to be a rural resident very much concerned with land ownership and property rights, wanting free rein to do as he or she pleases on his or her property, at times without regard for the consequences for future generations. The more conservative a person is the greater is the tendency to place primacy on humans of one's own race, creed, and religion, as well as on one's own personal needs, however they are perceived. The more conservative a person is the greater the tendency to disregard the land's sustainable capacity. Also, the greater the conservatism the more black-and-white one's thinking tends to be, which may have led American psychologist William James to observe that "a great many people think they are thinking when they are merely rearranging their prejudices."

A person with liberal thinking, on the other hand, is most often an urban dweller, who is likely to be concerned about the welfare of others, including those of the future and their non-human counterparts. Liberals also tend to be concerned with the health and welfare of planet Earth in the present for the future. And liberals more readily accept shades of grey in their thinking than do conservatives.

Conservative people tend to focus on individual pieces of a system, its perceived products, in isolation of the system itself, whereas liberals tend more toward a systems approach to thinking. A person oriented to seeing only the economically desirable pieces of a system seldom accepts that removing a perceived desirable or undesirable piece can or will negatively affect the system's productive capacity as a whole. Their response typically is "Show me; Prove it; I'll believe it when I see it." Resolution to environmental conflicts to such a person is usually seen as an immediate problem-solving exercise.

In contrast, a systems thinker sees the whole in each piece and is therefore concerned about tinkering willy nilly with the pieces because he or she knows such tinkering might inadvertently upset the desirable function of the system as a whole. A systems thinker is also likely to see himself or herself as an inseparable part of the system, whereas a piece thinker normally has himself or herself set apart from and above the system. And a system thinker is willing to focus on transcending the issue in whatever way is necessary to frame a vision that protects the system for the good of both the present and the future.

Unfortunately, when asking questions of value, we usually juxtapose conservative and liberal points of view, and then try to answer them with science. But science, which theoretically is the free pursuit of knowledge for its own sake, is the language of the intellect and was never designed to deal with values, which are the language of the heart. Both languages, however, are today necessary if human culture and its manifold environments are to be mutually sustainable.

LIMITATIONS OF SCIENCE

The great irony is that science, like values, is subjective. Nevertheless, the goal of scientific endeavour must remain the pursuit of pure knowledge, which demands unencumbered and open-ended inquiry while striving to be as objective as possible.

To keep the search for truth on its own credible track, one must first recognize the human tendency not only to form a single hypothesis but also to become so attached to it that any criticism of or challenge to one's methods raises one's ego defenses. This reflects the attachment to one's intellectual child, which is born the moment one derives what seems to be an original and satisfactory explanation for a phenomenon. And the more the explanation grows into a definite theory, the more attached to it one becomes. Then comes the massaging (as I've often heard it called in government agencies) of the theory to fit the data and of the data to fit the theory.

In addition, we tend to be "method-oriented" rather than "problem-oriented" in our thinking and therefore in many of the questions we ask. It's important to recognize this because we tend to think that through our experiments—our methods—we are learning the Truth about Nature when in fact we are learning only about our experimental designs—again, our methods—and our assumptions and expectations.

It is impossible to accurately "represent" nature through science, because scientific knowledge is not only a socially-negotiated, rigid construct but also a product of the personal lens through which a scientist peers. Scientists may attempt to detach themselves from nature and be "objective," but they are never completely successful. They are part of nature and must participate with nature in order to study nature.

As well, every scientist sees through his or her lens but dimly, first because we cannot detach ourselves from nature and second because all we can judge as fact are our own perceptions, which are always coloured by the lenses of our own value systems. Appearance, therefore, not reality, is all we can ever hope to see, and so it is appearance to which we often unknowingly direct our questions.

The secret about scientific research is that nothing can be proven—only disproven; nothing can be known—only unknown. So we can never "know" anything in terms of knowledge only in terms of intuition, which is the knowing beyond knowledge that is inadmissible as evidence in modern science. Whatever truth is, it can only be intuited and approached, never caught and pinned down.

VALUES AND SCIENCE

Over time, science, as imperfect as it is, has performed a vital function. It has piqued our imagination, challenged old ways of thinking, explored unknown phenomena, excited our sense of wonder and awe, elucidated relationships, and demonstrated the fuzziness of our world view. Science has shown that the sharp, clear lines of the mechanistic myth were derived by reducing ecological variables to economic constants. It has shown that the world in which we live is an interactive, interconnected, interdependent system in which the whole is expressed by the function of its parts, not by the parts themselves. And science has shown, albeit subtly, that there is no such thing as an independent variable, an interchangeable part, or a predictable outcome.

The salient point, therefore, is our illusion of definitive knowledge, not the state of our ignorance. And it is exactly because we are so certain of our knowledge that we are often so abysmally unaware of our ignorance.

Conservative thinking argues to retain the old reductionistic mechanical world view as its premise for decision-making. Liberal thinking argues for an evolving unified world view even though it is only now emerging into our consciousness. The conflict in decision-making, therefore, is between conservative and liberal value systems based on different world views, something science can address only indirectly.

The Endangered Species Act, for example, is first and foremost a question of values—do we or do we not as a moral act save our fellow species from extinction? Only secondarily is it a matter of science as we endeavour to save species in an attempt to understand their ecological functions and hence the environmental consequences of losing them—an understanding that may ultimately change social values.

In this sense, the Endangered Species Act is an ecological insurance policy for the future sustainability of human society within the context of a sustainable environment. But in its political enactment, it flows from long-term scientific inquiry back to immediate competing social values, to which the more conservative segment of humanity pays the most attention, to the potential detriment of environmental sustainability—present and future.

Sustainable Development of Forests as a Way to Preserve the Natural Basis of Forestry

Rainer Hummel
Alexander Sizykh

ABSTRACT. The term "sustainability" has been around for a long time and many definitions of the term are in use today. The concept has evolved over the years from meaning merely regeneration of trees, to a definition that considers all functions of the forest as being essential. However, valid doubt exists as to whether we could ever gather the extensive information needed to make accurate judgements about socioeconomic sustainability. The danger in too general a use of the concept is that it will become a blanket which covers underlying goals that are not compatible. Applied to forests, sustainability must include a consideration of preservation of genetic information to maximize the system's ability to respond to change. *[Article copies available for a fee from The Haworth Document Delivery Service: 1-800-342-9678. E-mail address: getinfo@haworth.com]*

AN HISTORICAL OVERVIEW

Sustainability has become a very common, even overworked term, that has gained general acceptance beyond the field of forest-

Rainer Hummel is affiliated with the World Forest Institute, Portland, Oregon, USA.

Alexander Sizykh is affiliated with the Russian Academy of Sciences, Siberian Branch/Institute of Geography, Irkutsk, Russia.

[Haworth co-indexing entry note]: "Sustainable Development of Forests as a Way to Preserve the Natural Basis of Forestry." Hummel, Rainer, and Alexander Sizykh. Co-published simultaneously in *Journal of Sustainable Forestry* (Food Products Press, an imprint of The Haworth Press, Inc.) Vol. 4, No. 3/4, 1997, pp. 53-60; and: *Sustainable Forests: Global Challenges and Local Solutions* (ed: O. Thomas Bouman, and David G. Brand) Food Products Press, an imprint of The Haworth Press, Inc., 1997, pp. 53-60. Single or multiple copies of this article are available for a fee from The Haworth Document Delivery Service [1-800-342-9678, 9:00 a.m. - 5:00 p.m. (EST). E-mail address: getinfo@haworth.com].

ry during the last decade. Since the Brundtland-Commission report was published (WCED 1987), the term has been used in many fields to describe development in accordance with nature. However, many definitions of sustainability are used simultaneously and the content is often not specified. One of the reasons for the various definitions is that the term "sustainability"–especially in German-speaking countries–was introduced a long time ago. It was first used in 1713 by V. Carlowitz, and it is not surprising that many different definitions have evolved over time.

Early definitions required that the harvest must not exceed the growth rate. Sustainability had a general meaning of steady, continuous yield and conditions opposite to terms like "exploitation" and "destruction." In the 19th century the understanding of sustainability covered only the forest area and aimed at maximum financial gain. Foresters only had to regenerate areas after harvesting in order to be sustainable (cut one, plant one–Judeich 1923). Since the 1950s, an increasing number of authors have considered all functions of forests as essential parts of sustainability including ecological and social aspects. Plochman (1982) for example has a holistic view that requires that forests be preserved as functioning systems. Speidel (1984) also includes all the services of forests and emphasizes the benefits for humans. A broader understanding of sustainability was also formulated in the Multiple-Use Sustained Yield Act in the United States in 1960. There, five categories of human benefits were identified: timber, fish and wildlife, outdoor recreation, range and fodder, and watershed protection. Glück (1994) points out that as a consequence of the Ministerial Conference on the Protection of Forests in Europe in Helsinki in 1993 the new concept of ecological sustainability was developed.

These extended definitions were developed simultaneously with increasing awareness that forests are not only a source of timber, but offer multiple benefits for the general public and the environment that should be maintained. Multiple-use forestry covers timber supply, ecosystem management, and recreation. The socioeconomic extension including ecological aspects, however, contains the danger that the term is only paid lip-service. Oesten (1993) doubts that all necessary factors of socioeconomic sustainability can be found and considered. It would require accurate information about all material and immaterial use of forests during a period of time. Bass (1993, quoted from Wiersum 1995), for example, lists the following data sets as crucial for the decision whether forests are sustainably managed or not.

Information on:

- balanced land use patterns: net changes into and out of forestry;
- maintenance of global ecological services such as biomass and net growth;
- sustainable management of forests: resource extent, size, condition and management intention and quality of its execution;
- forest resilience: ecosystem integrity (such as biological and structural diversity) and on the multiple uses of the forests;
- minimizing external inputs and waste: ecosystem capability and health;
- equity aspects and management accountability: forest ownership, use rights and legal status, actual use of forests.

Indeed, it is very unlikely that such extensive information can be obtained.

SUSTAINABILITY AND SOCIAL VALUES

Over time, the definitions of sustainability have always reflected the historical background and the prevailing economic theories. Most definitions have in common that the forest and its goods and services should be preserved for the generations to come. Different answers can be found to the question of what the effects and performances are that forests should sustain. Possible answers could be steady timber supply for the wood products industry, services for recreation and watershed or the conservation of forests as an ecosystem. As shown in the historical overview, sustainability has reflected in the past, and will reflect in the future, changing social values. The understanding of sustainability has been and will be an ongoing, dynamic process.

SUSTAINABILITY AS THE CULTURAL FOCUS
OF FORESTERS

Sustainability has gained broad acceptance over recent years as awareness of limited global resources has increased. The application of the principle is certainly a positive step, but there is also danger that the contents get blunted and generalized. Sustainability in this regard could become an icon for the good in general, covering underlying conflicts. The general acceptance can be misleading. It pretends harmony under the roof of sustainability, where many goals are summarized. Some of these goals are necessarily antagonistic.

A survey conducted among foresters in Germany shows how different the understanding of sustainability actually is (Schanz 1994). The survey reveals that all foresters associate the same connotative understanding of forestry, independent from age or position. "Continuation," "responsibility for future generations," "in the long term" and "progressive" are some of the terms most strongly associated with sustainability. Asked however, for the denotative understanding and the practical application, there are great differences depending on professional position and experience. For example, forest managers of the private sector view a stable economic situation as a prerequisite for sustainability, whereas foresters in the public sector put much more emphasis on public education and laws. These trends show that the common understanding of sustainability is mainly on an emotional level. The practical application requires specification and operationalization of the term. The results are in this regard coherent with the statement of Duerr and Duerr (1975) that sustainability is the focal point of faith in forest management. The fact that sustainability is mainly perceived on an emotional level contains difficulties for operationalization.

APPARENT HARMONY

This perception of harmony can conceal antagonistic goals and complicates discussion about sustainability. One model, introduced by Zonneveld (1990), about how to achieve sustainability is often quoted (e.g., Salwasser et al. 1993; Jensen et al. 1994).

According to this model the sustainable solution would fall in the intersection of the three spheres. Simultaneous integration of all factors provides optimal land-use alternatives. However, as a prerequisite condition, this model requires that there be an overlap. What happens if the ecological solution is not socially desirable? In this case, an iterative decision process is required to achieve the sustainable solution. Salwasser et al. (1993) point out that there is a high likelihood that desired conditions will not be sustainable if the balance among these three criteria is not reasonable (Figure 1).

Since the apparent harmony between the three circles is not necessarily given, contained goals have to be listed and ranked (Figure 2). Their interrelation has to be clarified. The assumption that the three fields have the same weight has to be questioned. The ecological capacity of forests must be the basis for social and economic demands. Human desires must take into consideration what nature is able to deliver. If conflicts occur, the ecological capacity must be the first priority if forests should sustain their performances for future generations.

ECOLOGICAL LIMITATIONS–SUSTAINABILITY AND ENERGETIC ASPECTS

Forests are ecosystems with complex and numerous interrelations among different species. The composition of these species

FIGURE 1. The sustainable solution as intersection of the three spheres.

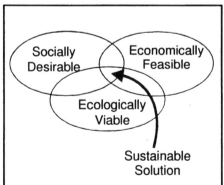

Source: Zonneveld 1990; Salwasser et al. 1993

FIGURE 2. The sustainability dilemma: the term sustainability can pretend non-existing harmony (Schanz 1994).

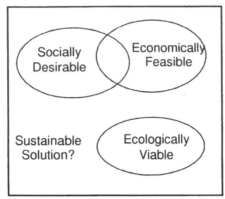

Source: Adapted from Schanz 1994

reflects the adaptation of plant communities to special site conditions and their genetic information. This adaptation and the resulting genetic information are a result of evolutionary processes. Natural systems are characterized by a minimal increase of entropy (Hinrichs and Schanz 1993). They have been adapted over very long periods of time and achieve a very efficient recycling of matter and energy in their system.

The entropy law (second law of thermodynamics), as stated by Clausius in 1865, says that the entropy of the universe tends to a maximum (e.g., Goldstein and Goldstein 1993). Thus the amount of energy available for work in a closed system only decreases. In other words, this means that any activity today is at the expense of potential activity for future generations. One of the conclusions of the entropy law is that perpetual-motion machines (100% sustainable systems) are impossible. On the other hand, sustainability, considering the entropy law, means to manage forests for a minimal increase of entropy. Prodan (1976) stated in the 1970s that the entropy law and its effects should be considered more in forestry.

Plant communities adapted over a long period of time have a minimal increase of entropy. Every modification of naturally adapted systems means necessarily a faster increase in entropy. Or, to put it another way, to achieve a degree of sustainability as high as possible, forests have to be managed as similarly as possible with naturally developed forests.

Considering this energetic aspect, the stability of vegetation communities and preservation of their genetic information should form the basis for sustainable forest management.

Preservation of all genetic information in forests includes another aspect of sustainability. Not only does it mean a minimal increase of entropy, but also a maximum capability to adjust to changes in the environment. In agriculture the loss of genetic diversity has already led to severe problems and shows how important the preservation of genetic diversity is, particularly when environmental conditions change. Managing species for maximum economic output may completely alter both phenotype and genotype. Considering that sites are subject to continuous change, both natural and man-made and man-made, capability to adjust to a changing environment could be another important aspect. Sustainability in this sense could be a comprehensive and long-term preservation of the natural basis of forestry in the face of challenges like the global climate change.

REFERENCES

Duerr, W.A. & J. B. Duerr. 1975. The role of faith in forest resource management. In: F. Rumsey & W.A. Duerr (eds.), Social sciences in forestry: A book of readings. W.B. Saunders, Philadelphia.

Glück, P. 1994. Enstehung eines internationalen Waldregimes. Establishment of an international forest regime. Centralblatt für das Gesamte Forstwesen. vol. 111, no.2, pp. 75-92.

Goldstein, M. & I. F. Goldstein. 1993. The refrigerator and the universe. Harvard University Press, Cambridge, MA.

Gowdy, J. M. 1993. Economic and biological aspects of genetic diversity. Society and Natural Resources, vol. 6, pp. 1-16.

Hinrichs, A. & H. Schanz. 1993. Forstliches Wirtschaften unter dem Aspekt der Entropieentstehung-ein Denkanstoß. Allgemeine Forst-und Jagdzeitung, vol. 164, no. 7, pp. 117-124.

Jensen, M. E., P. S. Bourgeron, P. F. Hessburg, R. L. Everett & B. T. Bormann. 1994. The process to generate and evaluate alternative sustainable ecosystem scenarios. Pacific Northwest Research Station–General Technical Report, vol. 1, pp. 7-10.

Judeich, F. 1923. Die Forsteinrichtung. Berlin.

Oesten, G. 1993: Anmerkungen zur Nachhaltigkeit als Leitbild für naturverträgliches Wirtschaften. Forstwissenschaftliches Centralblatt, vol. 112, pp. 313-319.

Plochmann, R. 1982. Der Forstmann vor der Herausforderung durch die wissenschaftlich-technische Welt. Der Deutsche Forstmann, 1982, H. 9+10, pp. 231-232, pp. 257-259.

Prodan, M. 1976. Verpflichtung der Forstwirtschaft und der Forstwissenschaft. Allgemeine Forstzeitschrift, no. 3, pp. 33-35.

Romm, J. 1993. Sustainable forestry, an adaptive social process. In: G. H. Aplet, N. Johnson, J. T. Olson & V. A. Sample (eds.), Defining sustainable forestry. Island Press, Washington, DC.

Salwasser, H., D.W. MacCleery & T.H.A. Snellgrove. 1993. An ecosystem perspective on sustainable forestry and new directions for the U.S. national forest system. In: G. H. Aplet, N. Johnson, J. T. Olson & V. A. Sample (eds.), Defining sustainable forestry. Island Press, Washington, DC.

Schanz. 1994. Forstliche Nachhaltigkeit aus der Sicht von Forstleuten der Bundesrepublik Deutschland. Working paper 19-94, Institut für Forstökonomie der Universität Freiburg.

Speidel, G. 1984. Forstliche Betriebswirtschaftslehre. Parey, Hamburg.

WCED (World Commission on Environment and Development) 1987: Our Common Future. Oxford University Press, London.

Wiebecke, C. & W. Peters 1984. Aspects of sustained yield history: Forest sustention as the principle of forestry–idea and reality. In: H. K. Steen (ed.), History of sustained yield forestry: a symposium. Conference Proceedings, IUFRO Forest History Group (S6.07).

Wiersum, K. F. 1995. 200 Years of sustainability in forestry: lessons from history. Environmental Management, vol. 19, no. 3, pp. 321-329.

Zonneveld, I. S. 1990. Scope and concepts of landscape ecology as an emerging science. In: I. S. Zonneveld & R. T. T. Forman (eds.): Changing landscapes: An ecological perspective. Springer-Verlag, New York.

PART TWO:
LOCAL SOLUTIONS

Prince Albert
Model Forest
Association Inc.

MODEL FOREST
NETWORK
RÉSEAU DE
FORÊTS MODÈLES

Challenges to Managing Forests in Africa

Mangetane Gerard Khalikane

ABSTRACT. Forests in Africa are being increasingly degraded because of increasing primary wood use and as a consequence of growing industrial activity. The challenges faced by African nations vis à vis natural resource use are many and include the need to work with local communities and to include them in the land tenure system, the need to regulate export of the benefits of natural resources and the need to provide alternative sources of income and energy sources. Meeting these and other challenges will require policies that ensure equitable distribution of benefits and of management rights. This, in turn, requires political stability and democratization. *[Article copies available for a fee from The Haworth Document Delivery Service: 1-800-342-9678. E-mail address: getinfo@haworth.com]*

NESDA

First let me take this opportunity to thank the organizers of this conference for inviting me to participate.

Not only in learning from you, but also in sharing my African experience with you.

The organization I work for is called Network for Environment, and Sustainable Development in Africa (NESDA). It is an Interna-

---

Mangetane Gerard Khalikane is Technical Advisor to the Network for Environment and Sustainable Development in Africa (NESDA).

[Haworth co-indexing entry note]: "Challenges to Managing Forests in Africa." Khalikane, Mangetane Gerard. Co-published simultaneously in *Journal of Sustainable Forestry* (Food Products Press, an imprint of The Haworth Press, Inc.) Vol. 4, No. 3/4, 1997, pp. 63-71; and: *Sustainable Forests: Global Challenges and Local Solutions* (ed: O. Thomas Bouman, and David G. Brand) Food Products Press, an imprint of The Haworth Press, Inc., 1997, pp. 63-71. Single or multiple copies of this article are available for a fee from The Haworth Document Delivery Service [1-800-342-9678, 9:00 a.m. - 5:00 p.m. (EST). E-mail address: getinfo@haworth.com].

tional NGO for African experts and their friends who are interested in fostering environmentally sustainable development. Ensuring sustainable management of forests and forestry resources is one of its many major goals. This can be achieved through partnerships between governments, private sector, NGOs and local communities. Put in a nutshell–ensure that all stakeholders are involved in management of natural resources.

NESDA's mission is to help African societies achieve environmentally sustainable development. Its strategies include:

a. Support to national planning processes for strategic sustainable frameworks like National Environmental Action Plans (NEAPs) and National Conservation Strategies (NCSs) through analysis of National policies and external reviews of "green plans."
b. Strengthening capacities of African countries to launch and implement strategic programs for sustainable development through support of experience sharing in workshops, seminars, and country visits to work with local professionals as counterparts in project preparation or studies.
c. Strengthening the network and information dissemination through creation of a roster of African experts in environment and development, foster increased cooperation among African countries, and facilitate access to environment and development information through INTERNET, publications and other means.

The NESDA brochure is available from the conference secretariat for those who are interested in knowing more about this International Non-Governmental Organization (INGO).

CHALLENGES

Now let me turn to the subject of this conference. In order to understand the challenges that are faced by Africa in managing forests, one has to follow a systematic approach in analyzing the major issues. The issues include:

a. What forests and where do they occur or simply where are they found?
b. What are their productive values?

c. Who owns them, or what tenure systems are involved in their allocation?
d. Who benefits from the management of forests?
e. How do you ensure fair distribution of these resources—including shares in benefits?
f. What should be the best management option or systems?
g. Do resources exist to support the best option? How can these resources be sustained?
h. Are there new ways of handling these issues (what can we learn from this conference)?

Ladies and gentlemen, let me now proceed to give brief responses to the questions that I asked a few minutes ago.

Where are these forests? Simply, they are all over Africa, except for a few countries such as Lesotho, where natural forests were never abundant. However, they have been lost in most countries through degradation and other causes. The Table 1 below shows

TABLE 1. Some Basic Statistics on the Central African Region (Based on: *Africa Today, 1991; **recent estimates of world deforestation)

Country*	Forest & Woodland Area (km²)*	Population (1989 estimates)*	Forest Cover (km²)*	Cultivated Area as Percent of Land Area*	Percentage of Economically Active People in Agriculture*	Annual rate of deforestation (ha)**
Cameroon	475,470	10,968,000	247,600	15.2	62.0	1,000,000
Central African Republic	622,980	2,840,000	358,200	3.2	63.7	5,000
Equatorial Guinea	28,051	430,000	12,950	8.2	56.5	22,000
Gabon	267,658	1,131,000	200,000	1.8	68.6	3,000
Congo	342,000	1,939,000	212,000	0.5	59.7	15,000
Zaire	2,345,410	34,846,000	1,759,600	3.5	66.3	400,000

only a few countries where most of the species of high world interest are still concentrated.

The major sources of forest depletion include: agriculture, mining and logging activities. These activities have to be controlled if the remaining forests are to be saved from further unsustainable deforestion. The rates of deforestation have been estimated by a number of organizations as reflected in Table 2.

Shifting cultivation practice is the greatest contributor to the rate of deforestation. Aggravated by population increase, it can accelerate the depletion of forests. The big challenge is whether production and conservation objectives can be reconciled in the management of forest resources. This is the subject which should be addressed by this conference. Other contributions to deforestation include commercial logging, fuel wood harvesting, and open cast mining.

Therefore, the second big challenge is to provide affordable, as well as accessible alternative sources of income and energy to the use of these resources.

What is the value of forests? Being foresters you are fully aware of the value of forests. However, let me emphasize a few that I consider very important for our African nations.

TABLE 2. Deforestation Rates for African Regions (Source: FAO, 1991; Groombridge, 1992)

Region	Number of Countries in Region	Land Area (km^2)	Area of Forests 1980 (km^2)	Area of Forests 1990 (km^2)	Deforestation Rate (%) 1981-1990
West Sahelian	8	5,280,000	419,000	380,000	0.9
East Sahelian	6	4,896,000	923,000	853,000	0.8
West Africa	8	2,032,000	552,000	434,000	2.1
Central Africa	7	4,064,000	2,301,000	2,154,000	0.6
Tropical Southern	10	5,579,000	2,177,000	2,063,000	0.5
African Islands	1	582,000	132,000	117,000	1.2

1. Forests support our national economies in a number of ways; food production, energy source, building materials, medicine, and habitat for wildlife.
2. Environmental conservation is a major issue that is also reliant on good management of forest resources. The forest role in this respect includes soil protection against erosion, purification of air and water, reduction of wind velocity, regulation of stream flow, global stability through hydrologic and carbon cycles and in influencing local and global climate change.
3. Biological diversity is enhanced by proper management of forests. We all know that the endangered species exist today because of overexploitation or loss of species genes and from deterioration of ecosystems. Most of the requisite habitat is offered by forests. Therefore, they play a great role in the preservation of biodiversity. An analysis of the threat to biodiversity highlights the following activities.
4. Ecotourism has become an important sector in many African countries. Most of the attractions for this activity are found in the forests where interesting flora and fauna are found. Moreover, some employment and income generation occurs as a valuable contribution to our economies.
5. The sum of the four activities above is income generation which is so essential for African countries. Forest resources generate a lot of income through sale of forest products, gene plasm, hunting rights, tourism fees and taxes, shares in foreign companies, and direct employment in logging, parks, craft firms, etc.

This analysis brings us to a third important challenge in managing forests in Africa. How do we ensure sustainability of income generation from forest resources? This calls for sustainable use of the forests through conservation and protection of biodiversity as well as environmentally benign ecotourism. All partners in forestry development, especially local communities, must receive the requisite education and training to be able to accomplish this noble goal.

Who owns the forests? This issue has been discussed in many fora. What is important for this meeting is that successful management of forests rests in the hands of those who are within and close to them.

Tenure systems have various forms that include government land, private sector and communal lands. Forests are also managed in these varied systems. However, it has been our lesson that reserves, parks and other forms of tenure systems that exclude local communities have very limited chances of sustained success. The use of guards does not pay, and poaching and resource destruction continue unabated.

On the contrary, where all stakeholders are involved in management and sharing of benefits, there are good signs of success. The idea is to enhance ownership and belonging which make communities feel obliged to protect the forest resource because it is for their well being. They then apply traditional methods of policing member behaviour and apply proper punitive action according to traditional and customary rules and principles. Where private holdings apply, there is need to encourage sale of shares to interested communities in order to enhance community interest.

Who benefits from forests? Up to now it has been governments and private foreign companies. Very few logging, tourism companies and gene-plasm banks are owned by Africans. Benefits are exported to Europe, eastern and western countries and are used for development activities overseas. One can call this resource erosion from Africa to west and east. Correcting this train of resources needs strong policies and laws to enforce partnerships between local communities, government and foreign companies as already explained in the foregoing paragraphs.

A fifth challenge relates to the ability and capability of African governments to resist uncontrolled forest resources and forest products export. Moreover, this may require policies that ensure that foreign companies give shares to local communities and/or commercial land and property rates under stringent compliance to environmentally sound sustainable management of forests.

How do we ensure equitable distribution of management of forests and their benefits? This is an issue which we are busy studying in many countries as we develop National Environmental Action plans or green plans. Many African countries are developing environment support programs which seek to empower local communities to participate fully in planning processes, decision making and implementation of natural resource management programs including forestry. The key to these efforts is political stability and democratization including decentralization of power, resources and institutions to the lower echelons of society. It is only when these processes are in place together with full government commitment that the partnerships that I am proposing can work.

Therefore, the sixth challenge is to foster and sustain democracy in order to facilitate participation of all stakeholders in decision making, planning and implementation of forestry programs. Given the fragile political systems of Africa, this task will not be easy. However, a gradual push towards involving local communities, private sector and NGOs should be encouraged by both locals and donors.

Mr. Chairperson, let me now finish by making proposals to cover the last two issues of concern as I listed them at the beginning of this brief presentation.

What is the best option for managing the forests? As foresters you may have the best technical options. However, our experience tells us that the best option lies somewhere with traditional systems that allow local communities to play a major role in decision making. Therefore the role of expertise in forest management should be to tap indigenous knowledge. In-build that knowledge in our modern systems and empower the local communities to take the lead in management of natural resources including forestry resources.

This takes us to our seventh challenge. Are we prepared to work with indigenous people who have traditional knowledge, skills, and cultural art for managing forests sustainably? We need this knowledge in order to complement our modern technologies. Forestry experts should engage in research and development that probes deep into indigenous knowledge of forest management if they wish to foster sustainable forest management in Africa. Build partnerships with traditional experts.

Do resources exist to support the new approach? Yes, if local communities are allowed to engage in fund raising activities, given shares where they are keen and interested, and donor funds are directed to support local initiatives instead of central government.

No, if the above issues are not resolved. But what is clear is that donor funds are not going to last. With the coming of interest to eastern Europe, Africa has seen more donor cuts in country allocations. Therefore, supporting local initiatives may be the right answer to future resource shortages.

Therefore, our challenge number eight is empowering local communities to raise resources for managing the forests. Are donors willing to channel funds directly to local community initiatives for fund raising or through local NGOs that are working with local groups? Governments have to encourage this new practice without undue reservations. Direct support of governments has not worked for the last 30 years, so it is about time to venture in a new approach.

Let me conclude by assuring you that local communities are equipped to rationalize an investment of money that can be raised by managing forest activities. They will not divide it amongst their families but instead they will plough it into useful community services such as water development and harnessing; energy development, including electrification of rural areas; communications; infrastructure including roads, schools, clinics and housing; education and health; and other pressing community services.

In summary, the challenges to sustainable management of forests are not restricted to, but include:

a. The big challenge is whether production and conservation objectives can be reconciled in the management of forest resources.

b. The second big challenge is to provide affordable, as well as accessible alternative sources of income and energy to use of forest resources.

c. Thirdly, how do we ensure sustainability of income generation from forest resources and ecotourism?

d. Fourthly, tenure systems that exclude local communities have very limited chances of sustained success—how do we include them?

e. A fifth challenge relates to the ability and capability of African governments to resist uncontrolled forest resources and forest products export.

f. The sixth challenge is to foster and sustain democracy in order to facilitate participation of all stakeholders in decision making, planning and implementation of forestry programs.

g. Our seventh challenge—are we prepared to work with indigenous people who have traditional knowledge, skills, and cultural art for managing forests sustainably?

h. Our challenge number eight is empowering local communities to raise resources for managing the forests.

i. And all other technical challenges of forestry management as well as use of appropriate technologies which you as experts in forestry management are vested with. So technical knowledge is our ninth challenge. To this end, we need training for all levels of our society. Moreover, good quality information for planning and decision making should be enhanced—it is our tenth big challenge in our African societies.

In conclusion, let me emphasize the need to ensure full participation and involvement of all stakeholders in forestry management. I quote the words of a convinced officer at TVA after going through a participatory review of planning activities under the authority "the various publics were apparently satisfied, not because each received everything requested, but because all requests had been heard and considered."

I once again thank you for this opportunity. Thank you very much.

REFERENCES

ATLAS/USAID: Natural Resource Management in Africa; Issues in conservation and socio-economic development, Liberville, Gabon, October 1993.

Carter, Nick: Environment–Forecasting Africa's Future, Africa Report 1990.

FAO: An Interim Report on the State of the Forest resource in developing countries, Rome 1988.

FAO: The forest resources assessment 1990 project; FAO's ongoing appraisal of the world's forest cover and recent trends in deforestation, Rome 1991.

Groombridge, Brian: Global biodiversity; status of the Earth's Living resources, World Conservation Monitoring Centre, Chapman and Hall, London, 1992.

IUCN, WRI & UNEP: Global Biodiversity Strategy, Policy Maker's Guide, 1992.

NESDA: National Conservation Strategies (NCSs) Reports; different countries in Africa, 1985-1994.

NESDA/World Bank: National Environmental Action Plans (NEAPs) Reports; several African Countries, 1989-1994.

Repetto, Robert: The Forest for the Trees; Government Policies and the misuse of forest resources, World Resources Institute, May 1988.

World Resources Institute: A Guide to Global Environment, Towards Sustainable Development, 1992-93 Report.

Community Forest Management in Southern Asia: A Survival Issue

J. B. Lal

ABSTRACT. Forest management is directed to attain one, or a combination, of the following three goals: stability of the physical environment, productivity of the physical environment and equity of the social environment. For the management to be sustainable, four aspects have to be addressed: the ecological, technical, the socio-economic and the institutional. All of these four aspects are equally important, and inadequacy of practices in any of the aspects could mar the sustainability of a forest. In India, where the rural population overwhelmingly depends on firewood for domestic energy, where the fodder for millions of cattle comes from forests, and where customs and traditions are as important in forest management as the considerations of ecological laws, a minimum of socio-economic data is a must for sustainable forest management. For India, this data must comprise:

- per capita annual requirement of firewood at a reasonably efficient level of utilization;
- carrying capacity of a forest unit in terms of its use as grazing land;
- per capita annual requirement of small timber for agricultural implements and for housing;
- per capita annual requirement of non-timber forest products (NTFPs);

J. B. Lal is Director of the Indian Institute of Forest Management, India.

[Haworth co-indexing entry note]: "Community Forest Management in Southern Asia: A Survival Issue." Lal, J. B. Co-published simultaneously in *Journal of Sustainable Forestry* (Food Products Press, an imprint of The Haworth Press, Inc.) Vol. 4, No. 3/4, 1997, pp. 73-88; and: *Sustainable Forests: Global Challenges and Local Solutions* (ed: O. Thomas Bouman, and David G. Brand) Food Products Press, an imprint of The Haworth Press, Inc., 1997, pp. 73-88. Single or multiple copies of this article are available for a fee from The Haworth Document Delivery Service [1-800-342-9678, 9:00 a.m. - 5:00 p.m. (EST). E-mail address: getinfo@haworth.com].

- customs and traditions prevalent in a community in terms of the use of forests;
- rate of growth of population, both human and cattle.

[Article copies available for a fee from The Haworth Document Delivery Service: 1-800-342-9678. E-mail address: getinfo@haworth.com]

INTRODUCTION: THE SHIFT IN FORESTRY

Prasana Khilari, a tribal elder of Munda community from the village Cingara in West Bengal, who has never been to school, does not know how to read and write, has never travelled by rail and does not have electricity at his house, says "Forests are like your eyes. You don't realize their value until they are gone" (Ford Foundation 1991). Khilari does not know what "economics" is, but he does know that a thing becomes a resource only when it gets scarce.

Katsura Watanabe, a forester herself, in a workshop organised in Nepal a couple of years ago, intelligently diagnosed "forester's syndrome," a common malady of foresters which results in their valuing trees over people and believing that the best trees are the tallest (Campbell 1993).

Khilari and Watanabe sum up the forestry situation in Asia. Many communities in Asia, poor and dependent on forests for their subsistence, came to realize that conservation of forests was not only an ecological issue, but a survival issue for them only after the deforestation rate had reached monstrous proportions in the nineteen-sixties. Unfortunately, they had not had any say in forest management and planning. The people to whom the forests mattered most, mattered little to those who managed forests. As Watanabe said trees were at the centre and people at the periphery.

In the last three decades, forestry in Asia has been marked by three major shifts: from sustained yield to sustainability, from timber doctrine to multiple use, and from the state dominance to people's participation.

The shifts were inevitable, and all would accept their validity, though some may do it grudgingly. There are, however, three crucial questions in regard to these shifts, and their answers must be found if the shifts are going to be meaningful. First, "what could be the operational definition of sustainability in regard to forest management?" Second, "what should be the parameters of multiple use?" We cannot expect the same package of economic goods and environmental series from every unit of forest. Should the forest be classified on a functional basis? Third, "what does constitute effective participation?" How should the people be represented in management bodies? Representative voices are, more often than not, weak. How should they be recognized?

SUSTAINABILITY

It is admitted that there cannot be a standard definition of sustainability. It would vary in the context of ecology, economy, and social and cultural environment, of the region. The definition also has to be flexible and innovative. Nevertheless, there has to be some kind of base definition which is less theoretical and more practical.

No South-Asian country has officially adopted a definition of sustainability in regard to forest use and management. Two definitions, however, appeal themselves, and could be useful in deciding if forestry practices are sustainable.

One was resolved in the Melesenki Conference of European Ministers in 1993. It reads: "use of forests and forest lands in a way that maintains their biodiversity, productivity, regeneration capacity and their ecological, economic and social functions while not causing damage to other ecosystems" (Lal 1994).

The other is the definition from Canada. Canada's goal is "to maintain and enhance the long term health of our forest ecosystems for the benefit of all living things, both nationally and globally, while providing economic, social and cultural opportunities for the benefits of present and future generations" (Forestry Canada 1992).

As a matter of fact, sustainability has three elements: ecological, economic and social. The two definitions take care of all the three elements, and so they are good. However, there are two points which are important in regard to sustainability but not explicitly mentioned in either of the definitions.

First, the integrity of forest ecosystems cannot be conceived in isolation. In a managed landscape, forests are very open systems. They influence, and in turn are influenced by other bio-productive systems, viz., croplands, grasslands and fisheries. To maintain the integrity of forest ecosystems, it is imperative that their management is integrated with those of the other three systems (Lal 1989). The research made at the International Centre for Integrated Mountain Development, Kathmadu, Nepal unequivocally brings out that the sustainability of production systems is inseparable from the sustainability of the resource base, and, to ensure the latter, an integrated approach in management of various bioproductive systems is called for.

The second of the two points is that the sustainability of social functions of forests depends on good institutional arrangement: the arrangements that can ensure equitable distribution of benefits from the resource (Lal 1994). It is the concern for the institutional dimension of forest management that has forced the governments, forest managers, and/or user communities themselves to evolve, the concept of 'Model Forest' in Canada, Joint Forest Management (JFM) in India, Community Forestry with user Groups (CF/UG) in Nepal, Forest Co-management in Thailand, Community Forestry Programme (CFP) and Integrated Social Forestry Programme (ISFP) in the Philippines, and Saemaul Undong in Korea.

INTEGRATED LAND USE–APATANI, THE UNIQUE TRIBE

The Apatani, a small but unique tribe in Arunachal Pradesh, a north-eastern state of India, needed no scientific research to adopt one of the most sensible land use systems in the world (Lal 1993).

The tribe confined to a small plateau lives in perpetual fear of a much bigger tribe, Nishi, who surround them. The reason for the fear is primitive. Apatani women are few of the most beautiful in the world, and the Nishi men would kidnap them if they got a chance. As a result the Apatanies do not move out much beyond their habitations. And, since, they would not move out, they grew their own forest and had their own grasslands and fisheries.

It would be no exaggeration to say that the tribe has evolved one of the best land use systems. They allocate level grounds and easy slopes to agriculture, steeper slopes to forestry, use marginal lands for growing grasses, and convert natural depressions into ponds for fisheries (Lal 1993).

The Apatanies grow only two forest species, one tree, and the other bamboo. The tree species is *Pinus excelsa*, and the bamboo, *Phyllosachys bambusoides*. The tribe has its own special technique of raising plantations and the technique incredibly ensures 100% survival (Lal 1993).

INSTITUTIONAL CHANGES IN FORESTRY: CANADA AND INDIA, A COMPARATIVE SITUATION

With more than 10% of the world's forests, Canada supports less than half a percent of the world's population. In contrast, with less then 2% of the world's forests India supports more than 16% of the world's population. Obviously, there are big differences in social and economic aspects of forestry in the two countries, but there are some similarities in institutional aspects.

The first similarity lies in the ownership of forests. Most of the forests are state-owned in both of the countries. The figures are almost identical: 94% of forests in Canada, and 95% in India belong to the state. The second similarity lies in national forest policies. In 1988, India resolved a new forest policy which contained a major shift from state-controlled, commercial forestry to forestry based on an ecosystems approach and people's participation. In 1992, Canada committed itself to a new national forest strategy which emphasized integrated and participatory forest management. In compliance to its new policy, India initiated JFM which made people the state's partner in forest protection and management, allowed full participation of key stakeholders, and enhanced science and technology development (Forestry Canada 1992; Lal 1989; Society of Promotion of Wastelands). Canada's Model Forest approach, however, is more balanced. India's JFM contains innovations only in institutional aspects. Model Forest acknowledges that technical innovations are equally important.

Perhaps, the similarities in the Indian and Canadian forestry situation end over here. There are differences in many respects, and these are big. Canada has over 40% of its land area under forest cover, and India has less then 20%. The figure for forest area per capita in Canada is an impressive 15.4 ha, and in India a pathetic 0.07 ha. In Canada, people mostly need forests for environmental services, recreation, aesthetics and industrial wood, in India, for subsistence, i.e., for food, firewood, and fodder. Of the total energy consumed in Canada, wood energy accounts for 7%, in India for 34%.

Canada is at the forefront of research on tree genes. The forestry research in Canada is at the level of cells; in India it is still mostly at the level of trees. Environmental ethics prompted Canada to plan Model Forests, but JFM in India is a survival issue. JFM is the only hope for conservation of forests and on forest conservation depends the subsistence of nearly one third of the people of India.

PARTICIPATORY MANAGEMENT—
THE INDIAN EXPERIENCE

Attempts at involving local communities in forest management is not a recent phenomenon in India, though as a rule, the country practised state-controlled, timber-oriented forestry which excluded people.

The first attempts were made as early as the nineteen-thirties in the Himalayan Uttar Pradesh, in the Bastar district of Madhya Pradesh, and in some parts of Himachal Pradesh, though they were motivated not by any ethical urge in the local governments, but by realization of the fact that effective forest protection cannot be achieved without the willing cooperation of local communities. The participation was limited too. The representatives of local communities only decided how the grazing in forests would be controlled and how the forests would be protected from illegal cuttings. The early attempts stayed local in nature and temporary in effect. The graduation of these local attempts to a national movement, took over fifty years. Indeed, it has taken over one hundred and twenty-five years for the people to find their rightful place in the forest management scene in the country. The following table (Table 1) indicates three distinct phases in forest management of the country.

The Government of India directives do not contain an official definition of JFM, but the words of Marcus Manech of Ford Foundation adequately define it. He says (Campbell 1993), "JFM is the sharing of products, responsibilities control and decision-making authority over forest lands between forest departments (of government) and local users' groups."

TABLE 1

Period	Philosophy of Forest Management
1860-1970	State-controlled, revenue oriented; production of timber primary concern.
1970-1990	Social and ecological considerations dominant in management; people's involvement negligible, though some successful experiments in forest management made in West Bengal, Haryana and Tamil Nadu.
1990 onwards	JFM becomes the offical policy with the issue of the Goverment of India directives in June 1990.

ARABARI EXPERIMENT

Joint Forest Management (JFM) which became agenda number one in forestry in India since the beginning of the nineteen-nineties started with a small experiment in Arabari, a small village in Midnapur district of West Bengal.

The Arabari experiment in people's participation was initiated as a result of repeated failures of forest functionaries to protect 'sal' (*Shorea robusta*) forests of a research centre by policing. Eventually, the Centre authorities sought people's cooperation in saving forests in return for the entitlement to usufruct and a share in final harvesting.

The pattern which evolved from the Arabari experiment and which has been officially adopted in the country, has the following key elements:

- People would be responsible for forest protection, and participate in management.
- They would be entitled to usufruct provided they organized themselves into a village institution.
- 25 to 80 percent (different states of the country have decided upon different norms) of the income from the final harvesting would go to the people.

A rather negative aspect of Indian JFM, however, is that it applies to degraded forests only. At every stage, government resolutions specifically mention that JFM extends to degraded forests only. Confining JFM to degraded forest gives the impression that governments don't have full faith in people, and don't want to risk the integrity of good forests by sharing their management with people.

JFM has two major lacunea also: first, no process has been evolved or prescribed to ascertain the majority opinion of the people, and secondly, the empowerment of weaker sections of the community has not been ensured. As a matter of fact, women still have little say in forest management decisions though they bear the brunt of the ill effects of forest degradation.

SUKHOMAJRI MODEL

Shivaliks located at the base of Himalayas are the hills which have suffered the worst of deforestation in India over a hundred years from the 1850s to the 1950s. The ugly face of soil erosion in the region was seen to be believed, before conservation efforts bore some results. On these hills, and in the catchment of Sukhna, the lake which provides water to Chandigarh, lies the village Sukhomajri.

Sukhomajri became suddenly important in the mid-nineteen-seventies when it was discovered that over two-thirds of the lake contained not water, but silt, and only good land practices could save the lake. Fortunately, the authorities realized that soil and water conservation could not succeed in Sukhomajri without making people equal partners in the effort. People's cooperation was sought and obtained. Participatory management of land resources including forests in Sukhomajri succeeded and Sukuna lake was saved.

Sukhomajri gave two lessons: first, people participate more enthusiastically, if forestry is not treated in isolation from other land activities such as agriculture, fodder production, etc., and, secondly, water conservation is given top priority. Therefore, if water is scarce in a locality, forestry should be primarily oriented to secure better water yields, quantitatively and qualitatively.

COMMUNITY FORESTRY
WITH USER GROUP (CF/UF) IN NEPAL

Nepal has had a unique forestry history. It was the last nation in Asia to bring forests under state control. The Forest Nationalisation

Act of Nepal came into force only in 1956. But with the Panchayat Raj Forestry Act of 1978, Nepal was also the first country to make legal provisions for people's participation in forest management. Under the provisions of Panchayat Forestry Act each Panchayat (or elected or nominated village body) was allocated 125 hectares of man-made forests, and 500 hectares of natural forests for management.

It was, however, soon realized that a Panchayat may not reflect true representation of primary users, and the gender bias may exclude representation of women altogether. Moreover, provisions of the Panchayat Act may not conform to customary forest practices. As a result, modifications were made in community forestry, and the system now known as CF/UG was evolved. In summary, the main processes of CF/UG are:

- to hand over accessible forest to communities;
- to empower user groups;
- to implement development and operational plans after approval by user groups; and
- to provide 100% of the products and incomes from forests to user groups (Campbell 1993).

The success of CF/UG prominently brings out two points: first, "social fencing" is more effective in protecting and regenerating forests than is the barbed-wire fencing and, secondly, the success of community forestry is inversely proportional to the difference between people's expectations and the actual accrual of benefits to them from community forestry.

FOREST COMANAGEMENT: THAILAND

Unlike India and Nepal, Thailand does not have a formal national police to enlist community support in forest conservation and management. But the fast declining ecology and economy in north-east Thailand forced the forest dwellers themselves into an effective forest conservation force. The regeneration of Dong Yai Forest in Ubon Ratchathani province in north-east Thailand presents an excellent example of the sustainability attained in use and conservation of natural resources with the partnership of user communities. What is more important and encouraging is the fact that the initiative for participation came from the communities themselves; the Royal Forest Department only responded positively.

Thailand suffered massive deforestation from the 1960s to the 1980s. In 1961, 54% of the area of the country was forested, by 1991 the forest cover in the country was reduced to half, and only 27% of the area remained forested. The reasons were two: first, unsustainable commercial logging, and second, large scale clearance of forests for agriculture.

Dong Yai Forest–Dong Yai means "big forest" in Thai–were cleared for cultivation of kenaf. Now these lands, though they supported big forests, could not sustain annual crops. Like most tropical forest soils, the fertility of Dong Yai soils is also dependent on their having a permanent forest cover. Kenaf cultivation failing, and forest gone, the already poor forest dwellers of Dong Yai were literally reduced to starvation. Forests, as sixty-year-old prom Chomchai of Dong Yai put it, give them rainfall, food and gum. Rainfall gave water, and gum secured cash (Poffenbergerm and McGean, eds. 1993).

As the investigations made by the Southeast Asia Sustainable Forest Management Network (Poffenbergerm and McGean, eds. 1993) reveal, Dong Yai forests yield over fifty edible leafy plants, thirty mushroom species, eight tuber varieties, fifteen fruits, and over twenty-five different edible fauna (squirrels, birds, ant eggs, lizards, snakes, fish, turtle, beetles, etc.).

The community concern and initiative and the positive response of the Royal Forest Department led to formation of village committees who prescribed rules for tree cutting, grazing regulations, and access to non-timber forest products, and made arrangements for protection of forests from fire and encroachers. Dong Yai forests are now regenerating fast, reviving the ecology and economy of the region.

The Dong Yai example proves four important points: first, high forest dependencies are compelling motivation to reverse forest degradation; second; a determined community can change the role of forest officials from authoritative custodians to that of educative and supportive partners; third, for success, community management does not need a written policy; and fourth; in addition to ecological and economic benefits, community management also yields moral benefits. The success of community management of Dong Yai forests has given forest dwellers a sense of confidence, self-reliance, and pride in their own capabilities.

INDONESIA: URGENT NEED
FOR COMMUNITY FOREST MANAGEMENT

Figures more than the words would tell the story of the fast dwindling forest resource in this country. The officially recorded forest area in the country is 144 million hectares, but only 109 million hectares are actually forested. Forests of adequate density extend over only 37 million hectares, the remaining forests are depleted to various degrees. Current estimates indicate that the deforestation rate in Indonesia may be as high as 1.2 million hectares annually. Half a million hectares of forest are lost on account of over-logging or poor logging practices, and the balance on account of agricultural conversions, fires and development projects.

Logging has earned billions of dollars for the national government, but it has brought few benefits to the forest communities. The overlogging has destroyed the economy of local communities who depended on traditional trade in such forest products as rattan, honey, aromatic wood, birds nests, resins and ironwood for shingles, for their livelihood.

Rehabilitation of degraded forest is an urgency both for the Indonesian government and for the forest communities. The government and the forest communities both cannot sustain their economy without well-stocked, well-regenerating forests. Government policies seek to regenerate urgently 20 million hectares of degraded forests, but as a former Environment Minister put it with the possible availability of funds, this target could be achieved only in sixty-five years.

The potential of Indonesia's forests to regenerate naturally, if adequately protected from fire and illegal fellings, is fully realized. The forest communities have both ecological knowledge based on generations of experience, and strong motivation to enable them to protect forests from further disturbances, and to let them regenerate naturally. All that is needed is that authorities give their cooperation.

As the studies made by the Southeast Asia Sustainable Forest Management Network in East Kalimanton reveal, the non-timber forest products from the Indonesian forests generate both cash and subsistence goods for forest communities. The communities would, however, harvest the products sustainably, and protect forests, if they are assured that they have exclusive and continuing rights to benefits. As the Network Report (Poffenbergerm and McGean, eds. 1993) puts it, ". . . . . given the limited field staff in the Ministry of Forestry, communities provide a major social and institutional resource for stabilising forest use."

PHILLIPINES:
INTEGRATED SOCIAL FORESTRY PROGRAMME (ISFP) AND COMMUNITY FORESTRY PROGRAMME (CFP)

In the 1940s, the Philippines extracted less than one million cubic meters of timber from its forests; in the 1970s it extracted over 15 million cubic meters. In 1990 it banned logging in remaining old growth forests (Poffenbergerm and McGean, eds. 1993).

There are parallel figures. In 1950, fifty-five percent of the land area of the country was under forests; in 1990, only twenty percent. During the period 1950-1988, the Philippines lost its forests at the rate of nearly 1,200,000 hectares annually. In 1990, the rate of deforestation came down to 80,000 hectares annually (Lal et al. 1994).

As the most of the forests were located on uplands, the people who suffered most on account of deforestation were the upland forest communities. The flow of forest products which formed their livelihood was reduced, and the productivity of their agricultural; land diminished as deforestation led to soil erosion.

Ironically, it was the forest uplanders who were blamed for the environmental tragedy. They were stigmatised as "slash and burn" farmers, and encroachers of public lands. The fact was that they moved from place to place as logging ruined their one abode after another.

In the 1980s, however, the trend changed. It was realized that if the forest communities were made allies, programmes could be successfully executed to achieve the twin objectives of poverty alleviation and forest rehabilitation. In 1986, the National Forestry Programme (NFP) based on ideals of social equity, people's participation and empowerment was initiated.

The programme has two major components, ISFP and CFP.

In ISFP, the NFP awards stewardship contracts to upland forest dwellers to authorize them to develop 3-5 hectares of land. It is mandatory to plant 20% of land with forest species; the remaining could be used for agriculture or rearing of livestock. As the contract is for 25 years, the communities do not suffer from tenure insecurity (Lal et al. 1994).

In CFP, the help of non-government organizations (NGOs) is sought to train communities in silvicultural practices. After the training, the NGOs and communities form partnerships to afforest degraded forest lands at the government's expense. After the plantations are established, i.e., three years after formation, NGOs withdraw, and the communities are awarded a twenty-five year contract to maintain and manage the afforested land. The communities get entitlement to usufruct and a percentage of earnings from the forests.

The NFP may not be a hundred percent successful—as a matter of fact, it is not, but, surely, it has created an environment in which people have started believing in their won capabilities and have begun to trust each other, and the government functionaries. With this attitudinal foundation, the success in community forestry is bound to come, sooner or later, especially as community forest management is one of the provisions of the new constitution of the country.

Korea: Saemaul Undong

The Korean forestry situation is different from that in the five Asian countries discussed before, in the sense that majority of forest lands—73% to be exact, are privately owned. The country has also made much more rapid economic growth than these countries. As a matter of fact, it is one of the few developed countries of Asia, and like any other developed country, it now relies little on biomass energy. But only until two decades ago it heavily relied on firewood for domestic energy and experienced acute firewood shortage in the 1960s.

Deforestation and population growth were the twin causes of firewood scarcity. To make up the firewood shortage, Korea launched the fuelwood plantation project in 1972. Significantly, Korea made the project a part of the Saemaul Undong, the national campaign to develop self-reliance, diligence and cooperation among rural population. Saemaul Undong aims at villagers themselves finding solutions to their problems through mutual discussion, selecting their leaders, and deciding upon the activities to be undertaken. Government's role is limited to initially providing financial, material and technical support.

After incorporation in Saemaul Undong, the responsibility of implementing the fuelwood project devolved upon Village Forestry Associations (VFA) which are grass-roots organization comprising small forest owners and villagers. The benefits accruing from the project are equitably shared by the forest owners and others participating in the project (Singh 1990).

As a result of the initiative, nearly 650,000 hectares of fuelwood plantations have been formed in Korea, which are not only producing firewood in plenty, but also secured many more social, environmental and moral benefits. Soil and water are conserved, agricultural productivity is enhanced, employment opportunities have been created and more than anything else, a spirit of self-reliance and cooperation has been generated in villagers (Rao et al., eds. 1994).

One more important lesson could be learnt from the Korean experience. As the economy of the country developed, the dependence on bio-energy reduced, rendering a big portion of firewood plantations economically non-viable. Had the plantations been a mix of timber and fuel species, they would have contributed more to the economy of the people, and of the country. The moral is that only the present should not be the basis of forestry decisions; the future should be envisaged, and the past reviewed.

CONCLUSIONS

Broadly, forestry serves three goals: stability and productivity of the physical environment, and equity of the social environment. It could serve all the goals in good measure only if all of its dimensions, viz., ecological, technical, socio-economic and institutional are adequately taken care of. Forestry, the world over, has suffered from one major draw back in the past: it gave emphasis only on one dimension at one point of time. In India, it was economic until the 1920s, technical until the 1960s, again economic in the 1970s, ecological in the 1980s and finally institutional in the 1990s. People's participation, in India, is now not only agenda number one, but agenda numbers two and three also.

Well, people's participation should be the agenda number one, but ecological, technical and economic issues cannot be ignored. Fortunately, Canada's Model Forest Program does not ignore any aspect.

The author's personal belief is that people's participation in forest management should not be treated as a means to an end, i.e., to check deforestation, or regenerate degraded forests. It is an end in itself. By any ethical criteria, people are supreme, and they should have a say in the management of a natural resource.

Forests influence not only other ecological systems, but also social, economic and political systems, and not only at local, regional or national level, but at a global scale. Caution and cooperation are the key elements in forestry, and should never be abandoned. "Caution" requires that research at all levels–landscape, ecosystem and species–should continue. "Cooperation" requires that partnerships be formed not only at local and national levels, but at international levels.

Forests do serve multiple functions, and should be so managed as to obtain multiple goods and services from them. Nevertheless, as stated earlier, we cannot expect the same package of goods and services from every unit of a country's forest. Functional classification of forests seems imperative to fix management priorities. We may note that, if not formally, then informally, forests both in Canada and India are functionally classified.

The famous ecologist Odum once said (Lal 1989) that an ecosystem approach to natural resource management lies in integrating ecology with economics. We would go a step further and say that environmental stability lies in integration of ecology, economics and ethnology. And people participation in forest management could achieve this integration.

REFERENCES

Campbell, J. G., J. Denholm (ed.) (1993). Inspiration in Community Forestry. ICIMOD Kathmandu.

Ford Foundation (1991). The Ford Foundation Letter. Vol. 212, No. 1.

Forestry Canada, Communication Division (1992). The State of Canada's Forests (1992).

Lal, J. B. (1989). India's Forests: Myth and Reality. Natraj Publishers, Dehradun.

Lal, J. B. (1993). Fifteen forestry principles: an approach to follow up. In: Lal, J. B., Uma Melkania, N. P. Melkania (eds.) Participatory Forest Management (Mimeograph), IIFM, Bhopal.

Lal, J. B. (1994). Forestry planning: New Challenges in Indian Forestry. A paper prepared for FAO.

Lal, J. B., Rekha Singhal and J. K. Das. Experiencing Community Forestry Programmes in Philippines: Modalities (1994) and Outcomes. IIFM, Bhopal.

Poffenbergerm and Betsy McGean (ed.) (1993). Research Network Report No. 2, 3 and 4. The Southeast Asia Sustainable Forest Management Network. University of California.

Rao, Y. S. et al. (ed.) (1994). Community Forestry, lessons from case studies in Asia Pacific region. FAO Regional office, Bangkok.

Singh, Samar (1990). People's Participation in Forest Management and the Role of NGOs and Voluntary Agencies. Paper presented to Seminar on Forestry for Sustainable Development Asian Development Bank. 7-8 Nov. 1990.

Society of Promotion of Wastelands. Joint Forest Management Regulations Update. SPWD, New Delhi.

Predivinsk Lespromkhoz: A Case Study on the Collaborative Restructuring of a Forest Product Based Community in Siberia

David Ostergren
Steve Hollenhorst

ABSTRACT. Ecologists, managers, federal forest officials, biologists, Russian and international environmentalists are developing and rethinking the practices and products of a timber operation in central Siberia. Predivinsk Lespromkhoz is a timber producing community located on the Einisei River, 150 kilometers north of Krasnoyarsk in central Siberia. The four thousand inhabitants are faced with expanding costs, a loss of financial support from the government, a collapsed market and disappearing natural resources. The restructuring of the Russian Government gave local foresters and timber operations more control of their resources but unfortunately, inflation and the dwindling economy restricted capital for implementation of new, sustainable forest practices and improved marketable products. Predivinsk is relatively accessible to foreigners inter-

---

David Ostergren is doctoral candidate in forest resources science, West Virginia University, USA.

Steve Hollenhorst is Associate Professor, Forestry, West Virginia University, USA.

[Haworth co-indexing entry note]: "Predivinsk Lespromkhoz: A Case Study on the Collaborative Restructuring of a Forest Product Based Community in Siberia." Ostergren, David, and Steve Hollenhorst. Co-published simultaneously in *Journal of Sustainable Forestry* (Food Products Press, an imprint of The Haworth Press, Inc.) Vol. 4, No. 3/4, 1997, pp. 89-102; and: *Sustainable Forests: Global Challenges and Local Solutions* (ed: O. Thomas Bouman, and David G. Brand) Food Products Press, an imprint of The Haworth Press, Inc., 1997, pp. 89-102. Single or multiple copies of this article are available for a fee from The Haworth Document Delivery Service [1-800-342-9678, 9:00 a.m. - 5:00 p.m. (EST). E-mail address: getinfo@haworth.com].

89

ested in the taiga of central Siberia and scientists from the Krasnoyarsk Forest Institute welcome collaborative projects. *[Article copies available for a fee from The Haworth Document Delivery Service: 1-800-342-9678. E-mail address: getinfo@haworth.com]*

INTRODUCTION

Each oblast, krai and republic within the Soviet Union had areas which were partitioned into a village-based system of forest management enterprise areas or *lespromkhoz*. The word "lespromkhoz" is derived from the three Russian words for forest, industry, and management ("lespromkhoz" is singular and "lespromkhozi" is plural). Using case study methodology, this paper describes the effects of the Soviet Union's 1991 demise on one of these enterprises, the Predivinsk Lespromkhoz which is located in central Siberia. We briefly outline the economic context of central Siberia, the forest resources of the Russian Federation, the Federal Forest Service, and the lespromkhoz system. We then describe the challenges and prospects facing the Predivinsk Lespromkhoz as it attempts to make the transition into the market economy. Our study is ongoing and these are the results thus far.

The case study approach allowed us to investigate the lespromkhoz phenomenon within its real-life context. The case study method was useful given our lack of control over the circumstances, the need to use multiple sources of evidence, and our interest in the intensive study of an individual lespromkhoz (Yin 1994, 1992; Platt 1992). The Predivinsk Lespromkhoz is under the influence of many factors which were incorporated into the analysis. These factors include its relationships with the government, private organizations, the academic community, the Federal Forest Service, and international and national non-governmental organizations. Additional factors are sources of capital investment, the personality of the director, the geographic location, and the needs of the people. The case study methodology enabled us to incorporate such information into our analysis.

We utilized several sources of information including, written records of the resources, production, and consumption of goods by the lespromkhoz. We also conducted informal interviews with participants and members of the community. Finally, passive observation of the community has been our largest source of information.

ECONOMIC CONTEXT

One reason for the USSR's demise was that the centrally planned economy failed to provide its citizens with enough quality consumer goods (McCauley 1993; Malarz & Roberts 1991). For decades, the command and control economy required that all decisions for production and delivery pass through planning ministries in Moscow. The ministries' job to plan production and anticipate demand was difficult, if not impossible. The production enterprises themselves were unable to respond to consumer demand and remained inefficient and wasteful (Nove 1992; Gregory & Stuart 1981). The system did not encourage innovation or individual initiative for problem solving by workers and technicians. The natural environment was considered a cost free source of raw materials and receptacle for waste products (Komorav 1981). Waste, bureaucratic politics, public concern for the environment, and finally glasnost contributed to the end of the Soviet Union.

The forest industry was not immune to the problems of the Soviet economy. Timber was extracted from Siberia and remote areas of European Russia and sold below the costs associated with its extraction and primary processing. Subsidizing the transportation system and wood products industry were an economic burden on society (Nove 1992). Resources were wasted by leaving harvested timber in the forest and losing timber during delivery between the harvesting site and the mill (Shvidenko & Nilsson 1994). Gusewelle (1992) recorded that a tremendous amount of raw timber was wasted during river transport as raw logs which floated down the Lena River were lost and finally washed up on the shores of the Arctic Ocean. In part because of this waste, the industry was unable to deliver enough building material and paper products to meet consumer demand. When the Soviet Union collapsed, all entities in the wood products industry faced an abrupt transition to the market economy with little or no government support.

SIBERIA

Arguably, Russia has the greatest amount of natural resources in the world (Pryde 1995:29). These resources include minerals, oil, gas, and hydroelectric power. In 1989-90 the former USSR accounted for approximately 26% of the world's forests (World Resources Institute 1994). Most of the forested area, in what is now the Russian Federation, is contained in the Siberian taiga. The taiga is the sub-arctic and arctic boreal forest that stretches from the Ural Mountains, across Siberia and the Far East, to the Arctic Ocean. The western Siberian plain is poorly drained and covered mostly with raised sphagnum-moss bogs. The Enisei River demarks the boundary to the eastern region that is hilly, with generally better drained soils interspaced with sphagnum-moss bogs. The main tree species include Siberian pine (*Pinus siberica*), the Scotch pine (*Pinus silvestris*), silver birch (*Betula pendula*), Siberian larch (*Larix siberica*) and various species of spruce (*Picea*), fir (*Abies*), and aspen (*Populus*) (Knystautas 1987; Holowacz 1985).

Historically, the vast wealth of Siberia has been mined and extracted to the benefit of European Russia. For 400 years, directions for the utilization of all resources in Siberia came from Moscow. Trees, furs, and minerals financed the development of Russia including Peter the Great's westernization of Russia, the late 19th century modernization of the army, and World War II (Barr & Braden 1988; Lincoln 1994). This tendency to view the region as a resource colony continues today. Containing 50% of the world's coniferous forests, international paper and softwood product companies view Siberia as a frontier for expansion while the Russians see the region as a source of hard currency (Backman & Waggener 1994; Shapiro 1994; Cardellichio et al. 1990).

Table 1 depicts the resources of the former USSR and Siberia and compares them to Canada and the United States. It is noteworthy that the United States has one third the forest area but had approximately 50% more roundwood production than the USSR in 1989-91. With vast, apparently untapped resources, the motivation to develop international logging enterprises is great. However, caution is suggested by analysts from both the environmental protection perspective (Gordon 1993; Grigoriev 1993; Rosencrantz & Scott

TABLE 1. Comparison of land areas, forested areas, and roundwood production.

Region	Land Area (thousand hectares)	1990 Natural Forested Land (thousand hectares)	1989-91 Roundwood Production (thousand cubic meters)
World	13,041,713	3,400,000	3,462,348
USSR (former)	2,190,071	754,958	375,400
Russian Federation	1,669,580	X	X
Siberia	1,180,000*	605,000*	125,600**
		(taiga) 450,000*	
United States	916,660	209,573	508,200
Canada	922,097	247,164	170,004

Source: World Resources Institute, 1994.
*Source: Shvidenko & Nilsson, 1994.
**1990 only.
X = unavailable.

1992) and forest utilization perspective (Shvidenko & Nilsson 1994; Mironov & Silachev 1993; Rilkov & Kotelnikov 1993; Braden 1991). The extremely slow growth rates and relative fragility of the taiga ecosystem are two major concerns that suggest the taiga does not recover well from intensive harvesting practices. Siberian forests are also threatened by forest fires, industrial pollution, and hydroelectric projects that flood productive forest lands (Scherbakova & Monroe 1995).

INCREASING PRESSURE
ON SIBERIAN TIMBER RESOURCES

The tendency has been to consider the distant resources in Siberia and the Far East region of Russia as inevitable areas of expansion. As a consequence, little effort has been put into regeneration in European Russia. In 1985-86 the cumulative European Russian backlog for forest regeneration (the difference between the forests planned for regeneration and those that have successfully regenerated) was 138 million hectares (Barr & Braden 1988). The lack of regenerated stands has contributed to a crisis in available timber and the subsequent increase in pressure to exploit the Siberian taiga.

In European Russia, economically accessible forests are now connected by road or rail. In Siberia, large areas are undeveloped and either can not be developed with foreseeable financial resources, or can only be developed with a great amount of capital investment. Backman and Waggener (1994: 16-17) contrast the *currently* accessible allowable annual cut and *potentially* accessible allowable annual cut for four regions of Russia. Forests are considered "potentially accessible" in the near term with considerable capital investment (see Table 2). Without increased regeneration efforts in European Russia the country will rely entirely upon Siberia and the Far East for future timber.

The domestic demand for forest products has been declining since 1990 and current prices are very low. These low domestic prices do not cover increasing harvesting and transportation costs. If and when the economy improves, prices will rise to cover costs and domestic demand and supply will increase (Backman and Waggener 1994). For industry located in central Siberia, long transportation routes to foreign markets also suggests their primary markets will be domestic. Furthermore, the lure of hard currency is prompting industry located closer to sea ports to respond to foreign demand. As timber is exported, there is potential for the creation of

TABLE 2. Currently and potentially allowable annual cut in Russia in millions of cubic meters (m^3)

Region	Currently Accessible	Potentially Accessible
European Russia	208	0
West Siberia	52	45
East Siberia (Predivinsk area)	109	57
Far East	57	17

Source: Backman and Waggener (1994: 16-17)

localized domestic timber shortages. These shortages will most likely be filled by central Siberian sources. Thus, the overall demand for Siberian timber resources is projected to increase because of a shortage in regenerated forests in European Russia, a lack of accessible forests to exploit, and an increase in domestic and foreign demand.

FEDERAL FOREST SERVICE AND REGULATIONS

The Forestry Committee is within the Ministry of Environment and Protection of Natural Resources. The Forestry Committee is the supervising body which comprises the Federal Forest Service. The Federal Forest Service adopted the Soviet system that divides the political regions of Russia into *leskhoz* (forest management) areas (Reidel 1992). These agencies guide the behaviour of industries utilizing the forest. In the past, all decisions came from the central government and individual decisions at the field level by leskhoz were rare. In March of 1993, the Russian Federal Forest Service was given a new set of laws (Soler-Sala 1994). The current transition in management represents a complete overhaul in the perspective and role of leskhozi and the Federal Forest Service.

Although the Russian government has retained ownership of all forests, management decisions are now initiated by the local leskhoz. Theoretically, these localized entities are better able to guide management practices and capture rent for the resources. An individual or corporate entity rents the right to use, for one to fifty years, the land for whatever economic purposes the corporate entity chooses (within the federal guidelines) which may include timber harvesting, hunting, turpentine production and/or preservation. The individual or corporate entity may be foreign or domestic. New and foreign operations are more likely to feel the effects of the 1993 forestry law.

The forested land of Russia is divided into three administrative categories (Pryde 1972:94; Backman & Waggener 1990: 44).

Class I: This category covers about 22% of the general forest land with the primary goal of retention/protection. Class I includes forests requiring the highest degree of protection such as national parks, urban green belts and parks, field and soil protective belts, health resort forests, forested belts along highways, railways and rivers, nut tree protection, and areas of regional significance such as whole mountain ranges. Only maintenance timber cutting with some thinning and small clearcuts are permitted in these forests.

Class II: This category is 6% of the general forest land. The category includes forests whose watersheds are of special significance, most of the lightly wooded areas in European Russia, and areas with a high density of transportation networks. Harvesting is allowed in these areas but the annual cut is not to exceed the annual growth. Frequently, the goal is expansion of the forested land. These lands are similar to national forests of the United States. Barr and Braden (1988) note that these forests have been reduced by over harvesting as a result of their proximity to transportation and population centers.

Class III: Industrial forests are 72% of the general forest land and 72% of the growing stock. This category includes all the densely forested areas of northern European Russia, the Urals, Siberia and the Far East. As mentioned earlier, much of this category has not been accessed and will be the focus of an expanded forest industry. These are commercial forests subject to intensive timber harvesting (Holowacz 1985).

THE LESPROMKHOZ SYSTEM

The forest product industry, including the lespromkhozi, was organized under the Ministry of Industry until the demise of the USSR. Currently, the entire system is shifting to the market system with the effect that ministries and working affiliations are changing from year to year. The government still influences the lespromkhozi but increasing numbers are becoming privatized alone, or with larger corporate entities.

From the period of 1917-1928, most forest utilization was similar to Tsarist times, in which the forest was primarily used by the peasant community for fuel, shelter, food, medicinals and other needs. Industrial harvesting activities were generally limited to regions near Leningrad mills for conversion to pulp and paper. Some timber was exported from European Russia to Europe for finished timber products (Backman & Waggener 1990; Barr & Braden 1988; Pryde 1972, 1991). Forest industries before World War II did not receive high priority for investment. Industrialization and national survival had precedence and as a consequence, investment in efficient harvesting techniques and regeneration was neglected.

Barr and Braden (1988) point out that by World War II, two thirds of the timber harvesting had shifted to the European north, the northwest, the centralized industrial area around Moscow, the Urals, and the Volga basin. They note 10% of the production came from Siberia and the Far East and that less than 1% of roundwood was exported for international trade. These data show that while timber was an important source of income for the USSR, most production was for domestic use. After World War II, the sale of raw materials such as timber (second only in importance to petroleum) helped finance the transfer of technology from the West to the USSR (Barr & Braden 1988). In this way, the foundation for today's Siberian forest sector was established.

Most of the lespromkhozi were temporary logging operations in the 1950s. As sawmills were built, these enterprises became permanent fixtures. Yet in many ways the lespromkhozi still reflect their original temporary character in that they lack permanent infrastructure such as sewage disposal or plumbing that can be found in most towns (Saltman 1994).

The lespromkhozi were instructed from Moscow as to their allowable annual cut and production goals. Often the cuts were measured in hectares harvested instead of product delivered. The resulting waste and overcutting left timber lying on the ground at landings and at mills. Indeed lespromkhozi were important to Moscow but not so important as to be utilized efficiently and updated to current world standards.

Currently, the lespromkhozi are facing a tremendous challenge. They still rely on the Class III forests, with established lespromkhozi utilizing their lands much as they did before the fall of the Soviet Union. In addition, they also must work with the leskhozi to negotiate long term contracts, follow federal management guidelines, develop markets, produce value-added products, and manage for the future facing true market prices for their product. The lespromkhozi in central Siberia lack fluid capital and have little prospect for domestic or foreign capital investment.

In many ways, the Predivinsk Lespromkhoz represents the problems and challenges of the Russian timber industry. They face the prevailing problems of the entire industry: lack of technology, outdated equipment, dwindling resources, inadequate transportation capabilities, and long distances to markets.

PREDIVINSK LESPROMKHOZ

Predivinsk is a lespromkhoz with a population of about 4000 people located on the Einisei River 150 kilometers north of Krasnoyarsk in Central Siberia. Built in 1928, their economy was centred around the construction of river barges. As the demand for barges declined they shifted to raw log harvesting and rough milling. Under the authority of the Ministry of Industry and the Committees of Logging and Wood Production, Predivinsk lespromkhoz was allocated approximately 250,000 hectares. Originally, lespromkhozi would relocate after exhausting their resources. Under the current policies, the Predivinsk lespromkhoz is expected to sustain itself with the original land base. The community lacks a central sewage or water system, paved roads, a method to pay for schools and other public services. Logging and wood processing are virtually the only industries, although many families depend upon the forest for hunting, gathering and garden plots.

The forest type is taiga (boreal) with Scotch pine, Siberian pine, spruce, fir, larch, and birch as the predominant commercial tree species. The only harvesting technique employed is clearcutting. The equipment used in harvesting is antiquated and heavy and has a severe negative impact on the soil and advanced regeneration. Like most lespromkhozi, Predivinsk does not replant the cut over forest with the more valuable coniferous species, instead allowing natural regeneration of the less valuable aspen and birch. Annually, the lespromkhoz cuts 3,000 ha from their allotment. They have already cut approximately 110,000 ha, leaving approximately 60,000 ha of commercial coniferous stands, with which to sustain their enterprise. At current cutting levels, the director estimates they will run out of forest resources in 10-15 years. That is to say the regeneration will not replace the rate of harvest and there will no longer be mature timber to harvest. The hauling distance to the mill is a costly 60 kilometers and is increasing as they access the remaining remote stands.

As do most lespromkhozi, Predivinsk would like to access the foreign market. The single largest barrier to those markets is the transportation distances. Timber is transported by barge upstream 150 km to Krasnoyarsk before it can be transferred to rail. From Krasnoyarsk, it is approximately 4000 km to either the Pacific rim or the Black Sea to access ports for further shipment. If milled lumber were to be shipped north via the Einisei River to the Arctic Ocean, the ice-free season is only three months long and the river distance is 1500-2000 km. The expense of transportation argues for value added products which can be priced to cover transportation costs.

Several individuals and organizations will be instrumental in the future of the Predivinsk lespromkhoz. The director—a combination of industry CEO and town mayor—is looking to new methods of developing his current resources. He is grooming his son (a Ph.D. in wood sciences) to build value added products for local domestic consumption in the urban areas such as Krasnoyarsk (pop. 1,000,000). However, the task is daunting without capital investment for new machinery and an ability to assess consumer demand.

The director is understandably reluctant to share information, accept others' opinions, solicit feedback from workers and in general, adopt new management techniques. However, he appears genuinely dedicated to the welfare of the community and is considering some of the new ideas proffered by foreign forestry experts and environmental organizations. The director is considering maintaining some of the area as undisturbed taiga hoping that the payoff from scientific and eco-tourism may be worthwhile.

Other entities are interested in helping Predivinsk in their transition while promoting models of sustainable forestry in Siberia. These include forest ecologists from the Russian Academy of Sciences, Russian and international non-governmental organizations, the USDA Forest Service, and Silva Forest Foundation, a Canadian nonprofit organization that does ecosystem based planning.

CONCLUSION

The major factors that will affect the future of the Predivinsk lespromkhoz include the limited amount of coniferous resource under their control, the slow regeneration and growth rate of boreal forest, the long hauling distances out of the forest and the prohibitive distances to markets. To make the transition into the Russian economy of the next century will depend upon the lespromkhoz's ability to identify and access markets and raise capital for modernization and reinvestment in new technology.

The new political and economic climate shifts responsibility for survival to the lespromkhoz. Questions arise as to options for the sustenance of their village and culture. Village leaders, government agencies, national and international interests, have invested a great deal of time and energy exploring opportunities for community development. Logging and primary wood processing will undoubtedly be a part of this picture. Value added processing also seems critical as they attempt to maximize the benefits of every square meter of wood before it leaves their control. Potential products include fiberization, particle board, boards, moldings and trim; but this all requires equipment modification or purchase. For this reason it is not in the village's interest to contract with domestic or foreign interests to supply raw, unprocessed logs. The village needs to make fundamental changes in harvesting and regeneration practices to ensure a continuous supply of coniferous stock.

Alternative economic development opportunities include various forms of international and domestic tourism such as ecotourism, cultural tourism, (folklore, rural life) science and educational tourism, visiting scientists, hunting and fishing tourism, and adventure travel. This would require investment in new village enterprises such as guide services, hospitality services, and transportation. Where the capital for these enterprises would come from is uncertain, but would likely include development banks for individual loans, market cooperatives and transportation cooperatives. A shift towards technology that is less expensive to purchase and maintain also appears to be a part of the equation.

It must be remembered that the forest is already an integral part of the informal economy of the village. Perhaps wild harvested products already utilized by the village can be developed into cottage industries. These include honey, aromatics, berries, mushrooms, medicinals, nuts, decorative wood botanicals, floral products, transplants, weaving and dyeing material, herbs, specialty wood products, management of wildlife for harvesting or sport hunting, agriculture, and agroforestry.

REFERENCES

Backman, C.A. & Waggener, T.R 1994. The Russian forestry sector outlook and export potential for unprocessed logs and primary forest products through 2000. Working paper #46, Center for International Trade in Forest Products, University of Washington.

Backman, C.A. & Waggener, T.R 1990. Soviet forests at the crossroads: emerging trends at a time of economic and political reform. Working paper #28, Center for International Trade in Forest Products, University of Washington.

Barr, B. & Braden, K. 1988. The disappearing Russian forest: a dilemma in Soviet resource management. Totowa, New Jersey: Rowman & Littlefield.

Braden, K. 1991. Managing Soviet forest resources. In P. Pryde (Ed.), Environmental management in the Soviet Union (pp. 112-135). New York: Cambridge University Press.

Cardellichio, P., Binkley, A. & Zausaev, V.K. 1990. Sawlog exports from the Soviet Far East. Journal of Forestry. June, 88(6), 12-16,36.

Gordon D. 1993. Russian Far East update. Taiga News. No. 5 March, 3-4.

Gregory, P.R. & Stuart, R.C. 1981. Soviet economic structure and performance. 2nd ed. New York, NY: Harper & Row.

Grigoriev, A. 1993. Leaving the door wide open for ruthless exploitation. Taiga News. No. 5 March, 2-3.

Gusewelle, C.W. 1992. Siberia on the brink. America Forests. 98 (5&6), 17-20.

Holowacz, J. 1985. Forests of the USSR. The Forestry Chronicle. October, 366-373.

Komarov, B. (pseudonym for Zeev Wolfson). 1981. Destruction of nature in the Soviet Union. Society 18(5) July/Aug., 39-50.

Knystautas, A. 1987. The natural history of the USSR. New York: McGraw-Hill. Book Company.

Lincoln, B. 1994. The conquest of a continent: Siberia and the Russians. New York, NY: Random House.

Malarz, A. & Roberts, I. 1991. Impact of political and economic changes in the Soviet Union on agricultural trade. Agriculture and Resources Quarterly 3(4), 527-42.

McCauley, M. 1993. The Soviet Union: 1917-1991. 2nd ed., New York: Longman Press.

Mironov, G.C. & Silachev, V.G. 1993. Ecology and new forestry methods (in Russian). Proceedings of the international conference of the socio-ecological problems of the Angara-Einisei River basin. Lesosibirsk, Russia, 20-23.

Nove, A. 1992. An economic history of the USSR. London: Penguin Books.

Platt, J. 1992. "Case study" in American methodological thought. Current Sociology. Vol 40(1) Spring, 17-48.

Pryde, P.R. 1972. Conservation in the Soviet Union. New York: Cambridge University Press.

Pryde, P.R. 1991. Environmental Management in the Soviet Union. New York: Cambridge University Press.

Pryde, P.R. (Ed.) 1995. Environmental Resources and Constraints in the Former Soviet Republics. Boulder, CO: Westview Press.

Reidel, C. 1992. Back to the future in the land of Genghis Khan. American Forests. 98(5&6), 21-24.

Rilkov, V.F. & Kotelnikov, A.M. 1993. Present day problems of the Eastern/Baikal forest in focus: ways of raising productivity and the need for new and more considerate forestry (in Russian). Proceedings of the international conference of the socio-ecological problems of the Angara-Einisei River basin. Lesosibirsk Russia.

Rosencrantz, A. & Scott, A. 1992. Siberia's threatened forests. Nature. 355, 293-294.

Saltman, S.M. 1994. Forest management in the Republic of Karelia. Journal of Forestry. 92(9), 37-40.

Scherbakova, A. & Monroe, S. 1995. The Urals and Siberia. In P. Pryde (Ed.) Environmental resources and constraints in the former Soviet Republics. Boulder, CO: Westview Press.

Shvidenko A. & Nilsson, S. 1994. What do we know about Siberian forests? Ambio. 23(7), 396-404.

Shapiro M. 1994. New Russia: a country on the take. The Washington Post. Sunday, November, 13.

Soler-Sala, P.A. 1994. Translation of the official text of the principles of the forestry legislation of the Russian Federation. Unpublished USDA Forest Service document. Washington DC. International Forestry Division.

World Resources Institute, 1994. World resources 1994-95: a report in collaboration with the United Nations Environment Programme and the United Nations Development Programme. New York: Oxford University Press.

Yin, R.K. 1994. Case study research: design and methods. 2nd ed. Thousand Oaks, CA; Sage Publications Inc.

Yin, R.K. 1992. The case study method as a tool for doing evaluation. Current Sociology. 40(1), 121-137.

Ecologically Sustainable
Land Use Planning
in the Russian Lake Baikal Region

Daniel R. Plumley

ABSTRACT. Efforts at achieving sustainable forest management throughout the world often appear to pit global economic and development interests against those who seek preservation and environmental protection of the Earth's resources. Such conflicts, where they do arise, are often unmindful of the full range of land use parameters which must be taken into account when developing sustainable forest models on a regional, if not subcontinental scale–the scale on which many multinational corporate entities now base their business planning. In fact, sustainable forest resource planning in the global marketplace must be integrated with regional sustainable land use, cultural, ecosystem and economic planning if the myriad forest attributes are to be maintained for future generations. The proposed

Daniel R. Plumley is Director of Natural Cultural Resources for the private not-for-profit, Ecologically Sustainable Development, Inc. For the past 3 years he has been undertaking sustainable land use planning and implementation projects in the Lake Baikal region of south central Siberia under the auspices of the Comprehensive Program of Sustainable Land Use Policies for the Russian Portion of the Lake Baikal Region. He is the first recipient of the Howard Zahniser Adirondack Award (October 1995) for his work in Russia seeking to establish the Oka Forest Preserve–a 4,000,000 hectare forest reserve in south central Siberia, modelled after New York State's constitutionally protected Forest Preserve in the Adirondack and Catskill Parks.

[Haworth co-indexing entry note]: "Ecologically Sustainable Land Use Planning in the Russian Lake Baikal Region." Plumley, Daniel R. Co-published simultaneously in *Journal of Sustainable Forestry* (Food Products Press, an imprint of The Haworth Press, Inc.) Vol. 4, No. 3/4, 1997, pp. 103-117; and: *Sustainable Forests: Global Challenges and Local Solutions* (ed: O. Thomas Bouman, and David G. Brand) Food Products Press, an imprint of The Haworth Press, Inc., 1997, pp. 103-117. Single or multiple copies of this article are available for a fee from The Haworth Document Delivery Service [1-800-342-9678, 9:00 a.m. - 5:00 p.m. (EST). E-mail address: getinfo@haworth.com].

103

paper would describe comprehensive sustainable land use planning being implemented in the Lake Baikal region of south central Siberia, Russia, which incorporates the full range of forest utilization from wild forest preserves to ecologically sustainable forest management for wood products. This paper describes similar projects being undertaken by Ecologically Sustainable Development, Inc. (ESD) in Mongolia, the Altai region of Russia, the Ussuri River Basin of Far Eastern Russia and China, Nicaragua and British Columbia. Finally, the authors make recommendations key to achieving sustainable forest policy on the federal, state, regional and local levels. *[Article copies available for a fee from The Haworth Document Delivery Service: 1-800-342-9678. E-mail address: getinfo@haworth.com]*

INTRODUCTION–WHAT BONE ARE YOU?

It is the tradition of the Buryat Peoples, an indigenous culture within the Lake Baikal region of south central Siberia, to address their elders with due respect. That tradition requires, in essence, after taking a long journey and seeking shelter in the home of an elder whom you may not know, providing the answer to a basic question, "what bone are you?" Or "from whom and where do you come?" Or "from which family and clan, or tribe of peoples do you originate?"

In addressing the subject of sustainable forestry, it is worthwhile to ask a similar question about our forests. From where have our forests come? And it must follow, if we are to speak of sustainability, to ask "where are our forests now?" And where are they going– or how will they be sustained in the future?

This paper will address some of these important questions in describing the implementation of the Comprehensive Program of Sustainable Land Use Policies for the Russian Portion of the Lake Baikal Watershed (hereafter called the *Comprehensive Program*) (Davis et al. 1993). First, this paper will present perspective on the critical connection between culture and sustaining our forests. Cultural context, often overlooked, is one of the most critical aspects associated with maintaining healthy forests. In culture, as with pure science, we can begin to answer the question, "where do our forests come from?" The paper will discuss known trends that establish a foundation for answering the question of "where are our forests now?" while giving perspective on where our forests may be headed.

The paper will conclude by describing briefly the attributes and process of the *Comprehensive Program*, as well as its implementation in the promotion of sustainable forestry and sustainable development in the Earth's oldest lake ecosystem, Lake Baikal of Siberia.

ON CULTURE AND LINKS TO SUSTAINING THE FOREST

Unlike 100 years ago, it is possible today to gauge the awesome historical relationship between human culture and natural resource protection, such as in the sustaining of healthy, productive forest ecosystems. Every human culture since the dawn of history (and often well before that) can now be assessed, at least to some degree, for the success to which it acted as stewards to its forest ecosystems. The researching of such cultural ecology begs side by side comparison and from such investigations much learning can be achieved.

Almost without question, researchers who undertake this interesting backward look reach the same conclusion that indigenous peoples developed a strong cultural basis on which their natural and forest resources had to be sustained into the future. Typically this involved direct mores, beliefs and, as in the Iroquois People, oral or written codes of conduct with regard to personal responsibility towards the forest and nature (Morgan, 1954).

The traditional Buryat and Soyot Peoples of Lake Baikal knew these fundamentals. Indigenous peoples throughout the world living in, or adjacent to, forests of all varieties have demonstrated an innate understanding of the need to sustain forests. They have done this in order to maintain the multitude of forest attributes which they provide, because they knew where the forests came from, where they were in the present and where they needed to be if the future was to be in harmony, whether it be one generation removed or seven. It is striking to learn that many of the indigenous peoples, or Earth peoples as they are often called, attribute their harmonious existence with nature as a direct result of their culture's close connection with their distant past.

SUSTAINABLE FORESTRY REQUIRES FORESTS SUSTAINED

In today's language, many native cultures assert responsibly that they are different than modern European cultures because they have not failed to remember their "original instructions" with regard to their relationship with the Earth. A culture without respect and understanding of such critical tradition is only "pop culture"–here today and gone tomorrow. Sustainable forestry without a forest first sustained is the same winsome fantasy. The process and product which we are acting on in the Lake Baikal region recognizes that sustainable forest management policy must be achieved within the context of a regional land use program that first, and foremost, sustains forests for all of its inherent qualities. While this ought to be common sense, modern experience, unfortunately, indicates otherwise.

In this age of microchips, space exploration and the internet, modern culture appears destined to relive past abuses of ignorance or outright greed towards the world's forests. It is well documented that sustaining forest attributes–from preserving biological diversi-ty, to carbon capture and ecosystem function, to the ability to pro-vide for recreation, fiber and wood products (and the myriad other forest benefits)–can not be accomplished by addressing site-specif-ic ecosystem management techniques alone. It is true the world's forests are degraded by poor site management. But more and more, the very forests we hope to one day manage sustainably are falling victim to the cumulative impact of a multitude of inordinate, uncon-nected and often, ill-advised land use decisions on the landscape at the local, regional, national and international level.

Most participants of this conference can acknowledge that the forest in the place where they grew up as a child is gone, altered or degraded. Many will recount the neighborhood forest that once sustained their youth and local wildlife only to find that now, years later, they are gone. These forests have been bulldozed and blud-geoned into nothingness, buried under tract housing development, or the asphalt of our exalted Wal Mart parking lots. This, more often than not takes place in outer suburban zones or rural areas as well as forested areas where local land use zoning has not been instituted. Such destruction continues day in and day out in small towns, scenic natural areas, mountain and lake communities and now, as often in developing countries. And it will continue into the future in, what will become, sadly, the lost "back yard" forests of even today's generation—unless people act.

THREATENING TRENDS
ON SUSTAINING OUR FORESTS

Much of the forest destruction, owing solely to the lack of comprehensive planning for the future on the local level, is matched by the aggressive nation state, corporate and multinational decisions that degrade or imperil forest sustainability on site and, indeed, the ability to sustain forests at all.

There are of course numerous examples nationally and internationally of imperiled forests including the loss of old growth forests in the Pacific Northwest and large tract forests throughout Canada (from cutting or hydropower development). In the northern forests of the Northeast United States, second home development is contributing to forest fragmentation and much has been written about the destruction of tropical forests throughout the world. Multinational corporations today often act seemingly unconnected to any nation state's laws or regulations while making decisions that direct the future of literally millions of hectares of forest land. This occurs due to poor forest ecological policy, such as committing investment towards the establishment and maintenance of southern yellow pine monoculture plantations in the US Southeast, or through destructive forest exploitation such as in the wholesale clearcutting of the virgin boreal forests of Far East Siberia, and elsewhere.

Despite the World's penchant over the past 30 years for the environment, the global trends depict clearly the loss of our forests. The following trends and effects have direct cultural, and socioeconomic links to forest destruction throughout the world:

- *Global Population Increases:* The population of the world, estimated at 5.7 billion in 1995 is expected to reach 10 billion, or double in size, by the year 2050 (United Nations Population Division 1992; Haub). It is expected that most of this population growth will take place in the developing world. Population growth often has a direct, prevailing impact in developing countries on forest ecosystem sustainability.

- *Annual Deforestation Rates*: Annual deforestation rates have been increasing mostly in tropical forest ecosystems, especially in the Asia and Pacific regions. Between 1980 and 1991 it has been estimated that tropical forests alone are being lost at a rate of 15.4 million hectares per year, or nearly 1 percent of the total tropical forest being lost annually (FAO 1993). Other forest types (boreal, hardwood, mixed forest, etc.) are also seeing increased rates of deforestation in different regions around the world.

- *Human Disturbance of Vegetated Land*: Russia and North and Central America are nearly equal in the amount of vegetated land (including forest) that has been subjected to medium and high rates of human disturbance, presently. Regions of Europe and Asia have received the most human disturbance of vegetated lands to present. Globally, at least 50 percent of the land can be classified as having either medium or high rates of human disturbance where either the carrying capacity of the land is being exceeded (by livestock for example), to the extreme where vegetative regrowth is limited, or impossible (Hannah unpublished data).

- *Consumption of Wood Products*: Viewing market consumption rates for different forest products can provide important insights to issues affecting forest sustainability. Roundwood consumption, for example, has seen the greatest increase between 1981 and present in the industrialized countries, not in the developing countries as might be expected (to close to 1.5 cubic meters roundwood per capita per year) (FAO, 1992). Forests in developing nations may come under increasing threat should they seek to keep pace with roundwood consumption rates of the industrial nations.

- *Effects on Forest Attributes*: Habitat destruction is one of the primary causal factors in extinctions; being the culprit in upwards of 36 percent of all established species extinctions to date (World Conservation Monitoring Centre 1992) It has been estimated that forest ecosystems provide home habitat for over 50 percent of the world's species, thus, realizing forest sustainability is imperative to biodiversity conservation and maintenance.

- *Historical Land Use and Cover Comparison*: Comparing historical continental land cover over time provides insight to the trends for forest sustainability. Clearly, the most significant change in land use around the world has taken place because of the reduction of forested areas and the increase or expansion of agricultural lands under crops (Richards, 1990).

The factors described briefly above, and others not mentioned here, will continue to provide important indices to forest sustainability as we enter the 21st century. It should become clear that proper planning to provide control over trends which negatively affect forests is paramount to our future.

SUSTAINABLE DEVELOPMENT AND LAKE BAIKAL

Based on the Bush-Yeltsin Agreement on Environment in 1990, ESD's President, George D. Davis, was invited to participate in an international conference looking at the future of Lake Baikal–the world's oldest, and one of its most unique, waterbodies. What he found at Lake Baikal has amazed the peoples of Russia and only recently, since perestroika, have Baikal's wonders become more well known to the western world, as follows:

- Lake Baikal is one of 4 major rift zone lakes in the world, occurring at the separation of tectonic plates. These plates are constantly moving apart, consequently deepening the lake which holds no less than 20 percent of the world's freshwater at any one time.
- Lake Baikal's depth reaches to over 1600 meters making it the deepest freshwater lake in the world. The lake's deepest recesses contain life forms which, in conjunction with Baikal's five miles of sediment, provide scientists with unheard of research opportunities into life on planet Earth.
- Lake Baikal is home to over 1800 endemic species. An incredible array of fish, shrimp, crustaceans, freshwater sponges and, even, the world's only freshwater seal, called Nerpa, reside at Baikal, and nowhere else on planet Earth.
- Lake Baikal is surrounded by arid steppe ecosystems, mountains ranging to over 13,000 feet, taiga forests of larch (tamarack), Siberian cedar (actually a 5 needled *Pinus* spp.), Siber-

ian birch, poplar or aspen species, alder, and numerous shrub and ground cover species, including incredibly diverse mountain tundras, mosses, fungi and lichens.

• Lake Baikal is home to several native peoples including the Buryat, Evenk and Soyot peoples and, ethnic Russians with heritage from all over the former Soviet Bloc countries. The lake is so large (over 400 km long by 40 wide), that the native people call it the Baikal Sea, even today.

At the request of a coalition of government officials, scientists, and NGOs in Russia's Buryat Republic, and Chita and Irkutsk Oblasts, a team of Russian and American scientists and public policy specialists with support from the MacArthur Foundation began work in 1991, using the Adirondack Park as a model, to develop a sustainable land use plan for the Lake Baikal watershed basin in Russia so as to avert the degradation of the lake's water quality from continuing point source pollution (two major paper plants, numerous factories and urban pollution from the lake's primary tributary, the Selenga River).

The study produced a vision for the future of Lake Baikal in the form of a policy document demonstrating the results of two years of interdisciplinary work and a generalized map (Figure 1) for the sustainable land use for the Russian portion of the Lake Baikal watershed which was published in March of 1993.

FOUNDATION IN NEW YORK'S ADIRONDACK PARK

The Lake Baikal *Comprehensive Program* has its foundation in the former lands of the Hodenosaunee Iroquois, where the first regional land use and zoning plan was adopted by law in New York's Adirondack Park. This mountain region of northern New York has seen many firsts in forest conservation and preservation. New York State established the Forest Preserve in 1885 in an effort to combat abusive timbering, forest fires and to secure the preservation of the area's key watershed attributes important to commerce. The State's Forest Preserve gained even further protection when the law became embodied within the protective covenant of the State's Constitution which holds, *"The lands of the State, now owned or*

FIGURE 1. Map showing study site location and Lake Baikal region.

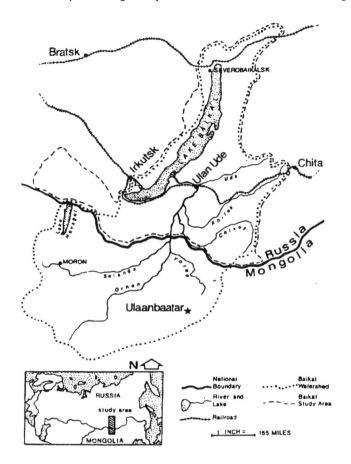

hereafter acquired constituting the Forest preserve, shall be forever kept as wild forest lands, nor shall the timber thereon ever be cut, sold, removed or destroyed."

In 1892, New York established the Adirondack Park, surrounding much of the Forest Preserve lands, and significant private holdings it had hoped to eventually acquire. This would not come to pass, however, and perhaps it is best for it enabled the park to become a proving ground for a modern day experiment of human habitation and communities living in harmony with forest wildness, albeit mostly second or third growth.

That experiment became tested in the 50s and 60s with the post-war development and recreational boom. Following a failed attempt at the creation of a smaller national park at its core, a study commission–the Temporary Commission on the Future of the Adirondacks–was formed by then Governor Nelson Rockefeller, who, two years later, would sign its recommendations into law as the Adirondack Park Land Use & Development Plan. The plan has been administered by the Adirondack Park Agency since 1973 and both its successes and challenges in implementation have provided a valuable model for the development of the *Comprehensive Program* at Lake Baikal.

LAKE BAIKAL: LISTEN TO THE LAND, LISTEN TO THE PEOPLE

In developing the Comprehensive Program of Sustainable Land Use Policies for Lake Baikal, the planning team led by George Davis (President, Ecologically Sustainable Development, Inc.) and Sergei Shhapkhaev (Director, Buryat Regional Department on Lake Baikal) developed a critical theme for such procedures: "Listen to the Land, Listen to the People."

"Listening to the Land" simply means evaluating, based on existing data and knowledge, the inherent capabilities and limitations of the land and water resources for various potential uses. All lands have inherent capabilities and limitations reflected by their soils, slopes, biologic communities, visual amenities and similar characteristics. Such factors can be evaluated individually and assessed or rated according to their ability to withstand sustainably or support certain land uses.

The process in Baikal, and in ESD's other program areas around the globe, undertakes a composite overlay mapping process whereby our cooperative land use planning teams make appropriate land use decisions by weighing these variable land characteristics in aggregate in conjunction with listening to the people.

"Listening to the People" describes the integral process of public involvement of stakeholders and the general public associated with the region in question. For Baikal this included 3 Russian states, the Irkutsk and Chita Oblast and the Republic of Buryatia. Through joint-Russian and American interdisciplinary planning teams, and public meetings to gain consensus and understanding on the program's goals and focus, specific land use decisions or futures can be determined, and considered as an added layer of information to factor into the composite overlay process (Davis et al., 1993).

THE LAKE BAIKAL GENERALIZED SUSTAINABLE LAND USE PLAN

The product of this extensive international work involving hundreds of Russian and American specialists included the establishment of a proposed regional land use plan and map for the 32,000,000 hectare Lake Baikal region in south central Siberia. The plan seeks to recognize the unique global nature of Lake Baikal through the designation of a proposed World Heritage Site Core Area, comprising some 21,444,000 hectares surrounded by a proposed World Heritage Site Buffer Area where more intensive human activity can take place in an area of roughly 11,340,000 hectares.

The land use plan established generalized zoning including 25 separate categories of land use which would best serve the people, while recognizing and best reflecting the limitations and carrying capacity of differing land types existing at Baikal. The land classifications range from agricultural lands for arable purposes, pasture and native hay lands, industrial lands, protected landscapes such as national parks and wildlife refuges to national scenic rivers and the region's critical airshed territory.

Each land use zone is further defined with its imperative purpose, policies and objectives and is complemented by both preferred and conditional land use listings. Such listings describe more clearly the types and extent of land uses appropriate for each zone. Preferred land uses are such land use activities that can be promoted with the individual zone. Conditional land uses require review and careful consideration within each zone and may warrant certain permit conditions by appropriate regulating bodies, or clear prohibition.

Finally, the zoning plan is complemented by performance standards that seek to better define preferred land use policies to be considered for all land use zones within the Baikal watershed. These performance standards attempt to define sustainable policies for agriculture, forestry, development and developed area use, industrial uses and mining, transportation and communication facilities, as well as lake, river and riparian area uses.

Paramount in the creation of the *Comprehensive Program* is the development of the following primary objectives for implementation of the Sustainable Land Use Plan for Lake Baikal:

1. To inextricably link sustainable economic development and environmental protection;
2. To preserve natural ecological processes and biological diversity;
3. To preserve cultural traditions and diversity;
4. To insure that present and future generations can live in dignity and will enjoy an improved quality of life;
5. To involve the public in land policy decisions;
6. To expand intergovernmental collaboration in sustainable development.

LAKE BAIKAL PLAN IMPLEMENTATION AND ACHIEVEMENTS

The Lake Baikal Sustainable Development Program has already demonstrated that both economic enhancement and environmental protection must be inextricably linked to preserve Lake Baikal. This will remain essential in protecting this watershed basin recognized internationally for its ecological importance, cultural diversity and vulnerability as a proposed World Heritage Site. Through 12 separate Model Projects funded through the US Agency for International Development (USAID), the recommendations of the *Comprehensive Program* are being realized by local and regional integrated land use planning, sustainable agriculture, sustainable forestry, international ecotourism, cultural preservation, wildlife restoration and the establishment of protected areas and sustainable economic enterprise.

The program has hastened the movement towards achieving regional land use zoning based on environmental and economic factors, that assure sustainability within the Lake Baikal Basin.

The 12 model projects are intended to demonstrate the United State's commitment to sustainable development, biodiversity preservation, democratic institution building, and Russian assistance directly to the regions, by:

a. Initiating small scale ecotourism, sustainable forestry, and sustainable agriculture projects consistent with the *Comprehensive Program*.
b. Initiating model projects to demonstrate the process by which the proposed comprehensive protected areas system recommended in the *Comprehensive Program* can be successfully implemented.
c. Providing technical assistance to the newly established Russian regional land use departments and other levels of Russian government necessary for the refinement, implementation and enforcement of the *Comprehensive Program*.

SUSTAINABLE FORESTRY AND FORESTS SUSTAINED

First and foremost, the *Comprehensive Program* recognized that unless we protect, maintain and preserve our forest ecosystems on the landscape level in the first place, sustainable forestry management will do little in the long-term to preserve critical forest attributes. Sustaining forest ecosystems is a primary result of the Comprehensive Program's land use plan through the designation for protection of 21,244,000 hectares of protected areas within the proposed Lake Baikal World Heritage Site Core Area.

These areas include scientific reserves (or Russian Zapovedniks), national parks, national wildlife refuges, national scenic rivers, protected landscapes, natural anthropological reserves and natural landmarks which would comprise approximately 65 percent of the Lake Baikal region. In these areas commercial forestry would not be permitted as the primary forest attributes of watershed and water quality protection, wildlife habitat, ecological and biodiversity conservation, sustenance of native peoples, wild land recreation and other values would be at the forefront.

Sustainable forestry management, as defined in the plan's performance standards and enhanced through new knowledge, is encouraged in the region's proposed World Heritage Site Buffer Area. This 11,340,000 hectare region is further delineated under various land use categories ranging from managed forest zones, limited production forests, watershed protection forests to rural settlement areas.

CONCLUSION

In conclusion, implementation of the *Comprehensive Program* has shown that by taking a holistic view towards our forest landscapes, by listening to the land, and listening to the people, we can meet the challenge of sustainable forestry. The promotion of the creation of the 4,000,000 hectare Oka Forest Preserve, modelled after NY's own Forest Preserve (described earlier) is but one case in point. Additional establishment of new protected areas, as well as the promotion of sustainable forest based industries and joint ventures in the Lake Baikal region through our sustainable forestry model project provide other cogent examples of sustainable development.

We believe this process is doable, and ESD, Inc. is furthering the cause of sustainable land use planning in Russia's Altai region, the Russian Far East and China, in the Lake Hubsgol-Selenga region of Mongolia, and will soon lead a similar effort in Nicaragua's Miskito Coast. Through these programs, cultures and people as widespread as America and Siberia can together gain needed insight into the key questions of where our forests have come from, where they are now and where they are headed. More important, these efforts can be replicated to enhance global knowledge and understanding of the interrelationship of forests and cultures, and through sound land use planning, permit our society to make wise decisions to insure a sustainable future of our forests.

REFERENCES

Davis, G. D. et al. 1993. *The Lake Baikal Region in the Twenty-First Century: A Model of Sustainable Development or Continued Degradation? A Comprehensive National Program of Sustainable Land Use Policies for the Russia Portion of the Lake Baikal Watershed.* A cooperative project of the Buryat Republic, Chita Oblast and Irkutsk Oblast lead by the Center for Citizen Initiatives

(USA), Center for Socio-Ecological Issues of the Baikal Region (Russia). New York: Davis Associates (USA), and the Russian Academy of Sciences.

Food and Agriculture Organization of the United Nations, *Forest Resources Assessment 1990: Tropical Countries*, FAO Forestry Paper 112, (FAO, Rome, 1993) as cited in Global Trends in Environment and Development, World Resources Institute.

Food and Agriculture Organization of the United Nations (FAO). 1992. *Agrostat PC on diskette*. As cited in Global Trends in Environment and Development, World Resources Institute.

Lee Hannah, unpublished data. (Conservation International, Washington, DC, 1993) as cited in Global Trends in Environment and Development, World Resources Institute.

Morgan, L. H. 1954. *League of the Ho-De-No Sau-Nee or Iroquois*. Reprinted by Human Relations Area Files, New Haven, Connecticut.

Richards, J. F. 1990. Land Transformation. In B. L. Turner et al., (eds.) *The Earth as Transformed by Human Action: Global & Regional Changes in the Biosphere Over the Past 300 Years*. Cambridge and New York: Cambridge University Press, p. 164, as cited in Global Trends in Environment and Development, World Resources Institute.

United Nations Population Division. 1992. *United Nations Population Division (UNDP), Long Range World Population Projections: Two Centuries of Population Growth*, 1950-2150 (UNDP), New York, 1992); and Carl Haub, Director of International Education, Population Reference Bureau, Washington, D.C. as cited in Global Trends in Environment and Development, World Resources Institute.

World Conservation Monitoring Centre. 1992. *Global Biodiversity: Status of the Earth's Living Resources* (London: Chapman and Hall), as cited in Global Trends in Environment and Development, World Resources Institute.

World Resources Institute. 1994. *Global Trends in Environment and Development*. A set of presentation graphs and maps illustrating major conditions and trends in the world's environment. 8 pages. World Resources Institute, Washington, D.C. 1994.

Forest Policy Evolution in Poland

Kazimierz Rykowski

ABSTRACT. In Europe, intensive forest management has anteceded a long history of intensive exploitation. The result has been the replacement of natural forests with highly productive albeit more vulnerable forests of simplified structure. Polish forestry has followed this same pattern and the country has seen a substantial increase in forest cover over the last 50 years. The main challenge for Polish forest management comes in trying to meet growing demand for wood products while responding to increased public demand for conservation of environmental values. Poland's new Forest Act for the first time puts environmental social and productive values of the forest on an equal footing. Within the framework of the Act, the Polish Policy of Sustainable Forest Development puts special emphasis on protection of biological diversity and the promotion of environmentally safe technologies and practices. The concept of Forest Promotion Areas is the major element of the Policy. The challenge for Forest Promotion Areas and for Model Forests will be to bring together all the necessary elements: environmental, economic and social that will allow sustainable development in its broadest sense. *[Article copies available for a fee from The Haworth Document Delivery Service: 1-800-342-9678. E-mail address: getinfo@haworth.com]*

HISTORICAL CONTEXT

Few people know that one of the most famous symbols of European culture, the baroque church of Santa Maria della Salute, which

Kazimierz Rykowski is Chairman of the European Forestry Commission and Deputy Director of the Polish Forest Research Institute, Poland.

[Haworth co-indexing entry note]: "Forest Policy Evolution in Poland." Rykowski, Kazimierz. Co-published simultaneously in *Journal of Sustainable Forestry* (Food Products Press, an imprint of The Haworth Press, Inc.) Vol. 4, No. 3/4, 1997, pp. 119-126; and: *Sustainable Forests: Global Challenges and Local Solutions* (ed: O. Thomas Bouman, and David G. Brand) Food Products Press, an imprint of The Haworth Press, Inc., 1997, pp. 119-126. Single or multiple copies of this article are available for a fee from The Haworth Document Delivery Service [1-800-342-9678, 9:00 a.m. - 5:00 p.m. (EST). E-mail address: getinfo@haworth.com].

decorates Canale Grande in Venice, was built on the trunks of 1,106,657 alder, oak, and larch driven into the mud of the lagoon.

This is only one of many examples of the price paid by forests for the development of our civilisation. We must still keep in mind the millions of hectares of forest harvested for diverse purposes: development of mining industry, architecture, building, trade, and of course for paper–the most widespread medium of that civilisation.

This reflection is to be an introduction to my contribution toward forest policy evolution in Poland. The European context as well as linkage with the past are necessary to assess today's Polish forest policy.

Over 30% of Europe is covered with temperate and boreal forests, and nearly all of them are man-made. This extensive reforestation followed the more or less intensive exploitation of forests in previous centuries on this densely populated continent.

We must remember that the concept of sustainable management was conceived in earlier times mainly as "sustainable yield," "sustainable wood supply," or "sustainable income" and was a concept originating in Europe several generations ago. The rule of forest sustainability was invented by Hartig (1804): "To guarantee for the future generations that they would be able to gain profit from the forest at least at the same level like in the present generation." The content of that rule appeared in the conception of sustainable development which was presented in the so called: "Brundtland Report" (1986), in the "Forest Principles" from Rio (1992), in the "Resolution H1" from the "Helsinki process" (1993) as well as in the "Montreal process."

Polish forest theory and practice were developing similarly as in other European countries, in conformity with that rule established at the beginning of the XIX century.

Let us hope that thanks to the recognition of Hartig's principle, European forests would withstand the most severe pressure from developing societies. But they have paid a high price: the previously natural and stable forest ecosystems were replaced in most cases by the contemporary highly productive forests of simplified structure and high susceptibility to damage.

The success of European forestry is until now of a quantitative nature, concerning especially forest cover, standing volume, mean volume increment, and relation between increment and harvest.

POLAND

Polish forestry is obviously in the mainstream of both negative and positive changes of European forestry. In 1945-1994 Poland increased its forest resources by more than 2.2 million hectares and it was the highest rate of the increase of the forest cover in Europe. We have now 8.7 million hectares under forests, it means 38% of the country's area. At the same time the growing stock increased from 906 million m^3 in 1945 to 1.502 million m^3 in 1993, and the average age–from 47 years in 1967 to 54 years in 1994. Thanks to those changes the forest harvest increased from 12.1 million m^3 in 1949 to 20.2 million m^3 in 1993, but not exceeding anywhere the current increment. The present annual harvest level (allowable cut) is about 60% of the annual increment.

We can add to that some positive changes in forest technology and in ecological character:

- the decrease of clear felling areas;
- the increase in naturally regenerated forest area;
- the increase in area of protective forests, preserving especially soil and water;
- the increase in area of national parks, and share of forest included into national parks;
- the increase in percentage proportion of broadleaved species in relation to the conifers.

However, the main threat endangering the continuing development of forests has come from the outside and has its source beyond forest management.

We have to restrain the phenomenon of the so-called "forest decline," consisting in the destruction of the environment as a result of the accumulation of the by-products of industrial development (sulphur, carbon dioxide, nitrogen oxides, fluorine, ozone and other products of photo-oxidation) in air, soil and water.

We have to face the growing threat and concern due to possible global changes, such as climatic change or depletion of the ozone layer.

We state that the list of species of flora and fauna endangered with extinction is still growing.

We observe still larger forest areas threatened by noxious insects and fungal diseases.

But it seems that the most important and crucial phenomenon which appeared around the world are the changes in social preferences and expectations concerning the forest. Still more emphasis is being put on the protection functions, environment forming and general public values of forests.

We must remember that the human population is growing at a rate of 92 million people per year.

In spite of the increasing Polish forest cover, the forest area per capita in Poland decreases continuously.

In this context we must be aware of wood consumption of the society, the same society which prefers the environmental value of forests to productive one.

Let us have a look at the amount of round timber harvested per capita in Europe, the world, and in Poland. It does not look like a situation that warrants optimism, especially if one considers the wood consumption per capita as an index of economic and social development.

We can say the same about paper consumption (Polish average consumption is 44 kg per capita per year).

World consumption of paper increases constantly and some prognoses say the current figure will double in the next 15-20 years.

Human dreams, wishes and needs often go in opposite directions. Man's expectations concerning the forest is an example of this. Will it be possible to manage all of them in the sustainable manner? If yes, how much time is needed?

Half a year before the Rio Conference, the new Act of Forests entered into force in Poland. The Act determined the rules of conservation, protection and enlargement of forest resources, and the rules of forest management in connection with other elements of the environment and the national economy. The new element of this law is connected with the changes of the previous hierarchy of the importance of forest functions and tasks: for the first time in this country and as the one of the first in Europe, the Polish forest legislation has equalled the environment, social and productive functions of forests.

Within the political and legal framework of "Forest Act" and "Environmental State Policy" the "Polish Policy of Sustainable Forest Development" was formulated in 1994.

The programme is composed of three parts: technological, educational and research. It considers three time horizons: the present day perspective, till the year 2000, and after the year 2000.

The main part of the programme corresponds to the technical-managerial documents "Guidelines of Forest Management," "Principles of Silviculture," and "Instruction of Forest Protection."

A special emphasis has been put on the protection of biological diversity and on the promotion of environmentally safe technologies.

Let me give you some examples:

- the forest ecosystem should be the object of forest management, and not the forest stand as was previously the case, and it should be analysed within the scale of larger spatial units;
- forest management methods ought to be individualised, taking into account ecological and land use regionalization;
- forest management plans should be supplemented with an annex on nature protection.

The programme formulates a variety of detailed recommendations and obligations:

- to preserve forest ponds, peat bogs, swamps, dunes, rock bassets, outcrops, and other so called "ecological values;"
- to leave in a stand 5-10% of old-growth trees till their physiological death, including also dead trees and trees with hollows as biotopes of many forest organisms;
- to limit the clearcuts down to 4 ha;
- to limit the use of chemicals.

The concept of establishment of "Forest Promotion Areas" is the most important element of the Polish Policy of Sustainable Forest Development. In December 1994 seven Forest Promotion Areas were established. They represent different natural conditions and they cover an area of about 5% of the total area of the state forests.

The goals of the Forest Promotion Areas contain introducing into practice, in co-operation with local communities, the principles of the forest management carried out in full recognition of the needs of nature conservation, in naturally coherent forest space being an element of a larger environmental system.

The main tasks were formulated as follows:

- comprehensive examination of the state of forest biocenoses and detection of their tendencies to change;
- maintaining or restoration of natural value of the forest, using methods of rational forest management based on environmental principles;
- integration of sustainable forest management goals with active nature conservation;
- promotion of multi-functional forestry and sustainable forest management including privatisation in forestry;
- carrying out forestry research and experiments;
- carrying out training for forest service administration and intensification of environmental education of the public.

The principles for protection and economic activity in the Forest Promotion Areas will be prepared by the Technical and Economical Commissions, consisting of representatives of forestry, wildlife sciences, biologists, scientific and technical organizations, conservationists and NGOs.

Approval and implementation of those principles will be carried out by public and scientific councils, designed by the Minister of Environmental Protection, Natural Resources and Forestry.

The first and the most important Forest Promotion Area is the most valuable ecosystem in Europe, maintained in a state close to nature–Bialowieza Primeval Forest. The special rules of the most rigorous form, for protection and management were approved by Minister of Environmental Protection, Natural Resources and Forestry at the end of 1994. These particularly strict conservation measures will make it a reality that half of the Bialowieza Primeval Forest–about 25000–hectares becomes a nature reserve according to IUCN standards.

The Bialowieza Primeval Forest occupies a particular position in the programme of the Polish Policy of Sustainable Forest Management–it is a model of coexistence between the socially useful many-sided functions of forest and the persistence of nature.

Also for the European and world forestry it is an unique chance to improve the sustainable forestry with the effective action of nature, both in the Bialowieza National Park and in forest reserves that were previously established and that are being currently approved.

Nevertheless, instead of co-operation and partnerships the foresters are attacked and accused by loudly and emotionally expressed opinions of the over-exploitation, destruction and annihilation of "human heritage" in Bialowieza Primeval Forest. Unfortunately, these opinions are disseminated abroad, also by internet, to mobilise the international public opinion. Sometimes it is a kind of eco-hobby and there is nothing that can be done about it.

But, if anybody encounters this kind of message, please be so kind before taking your position to at least check the information on statistical data.

Before I finish my contribution let me go back to some general statements.

The strategy of sustainable development in its broadest sense needs the following:

- a political system that would ensure the effective participation of citizens in decision-making;
- an economic system able to create a production surplus and technical high-tech know-how in a lasting and independent way;
- a production system that would respect the ecological fundamentals of development,
- a social system able to mitigate tensions resulting from the non-harmonic development:
- a market system that would favour the stabilisation of the structure of trade and finances;
- an administration system that is flexible and able to self-correcting.

These indispensable conditions of sustainable development are rather rare at the same time and in the same place.

But this is exactly a big challenge not only for the forestry and for the conception of "Model Forests" or "Forest Promotion Areas," but for whole national economy of each country.

Integrating a Research Station into Community Development and Area Protection in Nicaragua

Hans G. Schabel

ABSTRACT. The entire southeast of Nicaragua and parts of neighbouring Costa Rica are presently being organized according to an ambitious regional development scheme called Si-a-Paz (Yes-to-Peace). Included are areas from high to low management intensity, such as sustainable agricultural and ranching developments, forest restoration and buffer zones, cultural monuments, as well as biological and wildlife reserves. Integrated in this grand scheme is the isolated subsistence community of New Greytown (San Juan del Norte), access point to a stunning variety of unique habitats, including riverine, estuarine, marine, coastal lowland, and upland environments, which are virtually intact at this time. A research station is planned for this growing community, with the intent to not only serve the exploration of the nearby Rio Indio-Maiz Biological Reserve, but also to contribute to the economic development of the community itself, by assuring scientifically sound resource management input, diversification of the local economy and environmental education of the public. This paper focuses on the history of this project, and the interaction of governmental and non-governmental organizations in making it happen. *[Article copies available for a fee from The Haworth Document Delivery Service: 1-800-342-9678. E-mail address: getinfo@haworth.com]*

Hans G. Schabel is Professor of Forest Protection and Director of International Resource Management, College of Natural Resources, University of Wisconsin, USA.

[Haworth co-indexing entry note]: "Integrating a Research Station into Community Development and Area Protection in Nicaragua." Schabel, Hans G. Co-published simultaneously in *Journal of Sustainable Forestry* (Food Products Press, an imprint of The Haworth Press, Inc.) Vol. 4, No. 3/4, 1997, pp. 127-138; and: *Sustainable Forests: Global Challenges and Local Solutions* (ed: O. Thomas Bouman, and David G. Brand) Food Products Press, an imprint of The Haworth Press, Inc., 1997, pp. 127-138. Single or multiple copies of this article are available for a fee from The Haworth Document Delivery Service [1-800-342-9678, 9:00 a.m. - 5:00 p.m. (EST). E-mail address: getinfo@haworth.com].

INTRODUCTION

In the 1980s, the cooperative development emphasis between industrialized and less developed countries experienced major shifts, in part because the "trickle-through" theory for poverty alleviation in developing countries had not yielded expected results. In many cases, both economic and environmental indicators verified declining living standards and accelerating trends in environmental degradation.

A new concept began to take shape. The Brundtland Report, a blueprint for "sustainable development," envisioned a balance between economic and environmental concerns globally (World Commission on Environment and Development 1987). Concurrent with the emergence of this emphasis occurred the widespread decentralization of political control in many countries, and a more regional and community-based, i.e., a participatory or bottom-up approach to rural development.

This paper reports on one case study from Greytown (alias San Juan del Norte), Nicaragua, where attempts are underway to develop a research station, which will contribute not only to scientific knowledge, but also to greater economic stability in the community itself and, thereby, to the preservation of the area's rich biodiversity.

BACKGROUND

National Context

Nicaragua is Central America's largest and resource-richest country. Presently with the region's second lowest population density, its demographic structure and annual growth rate of 2.7% nevertheless mean significantly more people in the years ahead. Because of fertile volcanic soils in the western and northern part of the country, most of Nicaragua's population is concentrated there, while the heavily forested East, always with sparse and even declining human population during the recent Civil War, now faces a significant wave of landless colonists, drawn to the region's unspoiled natural resources (Wille 1991). Unlike the West, where ranching and cash crops (cotton, sugarcane and coffee) provide the economic base, soils and the humid climate beyond the agricultural frontier of this eastern lowland region, do not generally favor infrastructure development, and commercial farming or ranching are of limited potential at best. As a result, people in this region generally make a living by extracting natural resources subsistence-style.

During the last 15 years, Nicaragua has experienced three vastly different governments, the latest, a coalition of 14 parties having been elected in 1990. The aftershocks of a decade of civil war, adjustments to the new democratic society, and the complications of an indebted nation trying to function in an increasingly global economy, however, continue to seriously impede progress. While various political factions are still debating an array of socio-political issues, they agree at least in principle (and in line with the constitution), on the fundamental role of natural resources in the recovery and future well-being of the nation and the region at large. This common goal found expression in the recently released Forestry (PAF-NIC) and Environmental Action (PAA-NIC) Plans, detailed blueprints for sustainable development in Nicaragua (Anon. 1993).

Regional and International Context

One of the emphases outlined by PAF-NIC, concerns itself with the "conservation of forest ecosystems and biodiversity," and more specifically identifies the International System of Protected Areas For Peace (Si-a-Paz) as one of three priority initiatives for the conservation of biodiversity in Nicaragua and neighbouring Costa Rica. Si-a-Paz constitutes an international biosphere reserve, which attempts to territorially organize the entire southeast of Nicaragua and neighbouring areas of northeastern Costa Rica, in a coordinated mosaic of areas of different management intensity and purpose (Anon. 1988). This includes a zone for development of sustainable ranching and agriculture, a buffer zone for sustainable forestry, and a set of other areas to be conserved for their natural and cultural assets, including a national park, Indian reserves, national monuments, wildlife refuges, and biological reserves (Figure 1). As a binational, transborder project, Si-a-Paz attempts to safeguard the political environment of this permanent tension zone (Girot and Nietschmann 1992), by developing a viable economy, and thereby

FIGURE 1. International System of Protected Areas for Peace (Si-a-Paz), located in southeastern Nicaragua and northeastern Costa Rica (Modified after Anon. 1991).

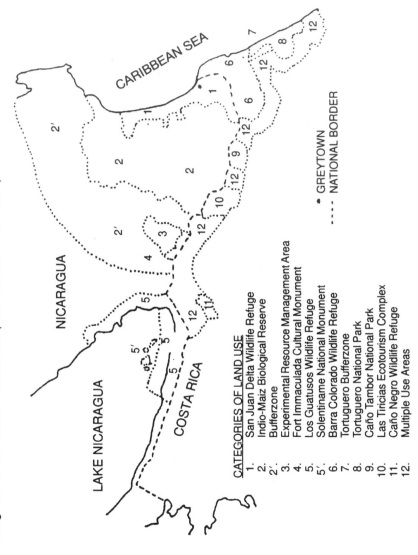

CATEGORIES OF LAND USE
1.   San Juan Delta Wildlife Refuge
2.   Indio-Maiz Biological Reserve
2'. Bufferzone
3.   Experimental Resource Management Area
4.   Fort Immaculada Cultural Monument
5.   Los Guatusos Wildlife Refuge
5'. Solentiname National Monument
6.   Barra Colorado Wildlife Refuge
7.   Tortuguero Bufferzone
8.   Tortuguero National Park
9.   Caño Tambor National Park
10. Las Tricias Ecotourism Complex
11. Caño Negro Wildlife Refuge
12. Multiple Use Areas

* GREYTOWN
- - - - NATIONAL BORDER

save its unique natural endowment, i.e., balance production and development with conservation in what is considered the most important conservation and development program in Nicaragua and Costa Rica (Anon. 1991).

On a larger scale, Si-a-Paz is envisioned to become a major link in an even grander scheme called Paseo Pantera (Path of the Panther). This multinational plan concerns itself with retaining and restoring as many natural areas as possible in a continuous vegetative corridor along the Caribbean coast, reaching from Columbia to the Yucatan (Marynowski 1992). Without it, the increasing fragmentation of forests is feared to disrupt the intercontinental exchange of biota from North and South America, which has contributed to the extraordinary exuberance of life on the isthmus, one of the world's motherlodes of biodiversity.

Greytown and the Indio-Maiz Reserve

In the remote southeast of Nicaragua, near the mouth of the Rio San Juan, the border river with Costa Rica, lies the community of Greytown. Its strategic location controlling passage to the interoceanic route, was recognized as early as 1536, and dreams of a canal surfaced with amazing persistence (the latest in the late 1980s), contributing greatly to an extraordinary history of booms and busts (Brannstrom 1992). From here pirates, filibusters and other desperados of various nationalities entered the river in endless attempts to gain access to the Pacific, or to ransack El Castillo, San Carlos and Granada. The British in particular, most notable among them Admiral Nelson, tried to gain control of the river, its fortifications, and the lake in order to split up and dominate the Spanish Americas. San Juan del Norte was taken and fortified by the British repeatedly in the 1700s. After abandoning it briefly in 1786, a contingent of English and Miskito Indians recaptured the port in the name of the Miskito king. After yet another British occupation in 1847, the town was renamed Greytown, after Jamaica's governor Sir Charles Grey. From there on the Americans increasingly joined the fray.

The California gold rush starting in 1848 initiated a major boom for Greytown. Vanderbilt's Transit Company controlled the transfer of prospectors arriving by ship from the Eastern U.S., up the river, across Lake Nicaragua, to the Pacific. From the 1860s to 1890s, Greytown reached its commercial pinnacle. For a number of years, part of the coast was declared a British protectorate, with even a consul resident there. In the words of Belt (1874), Greytown "was then a magnificent port . . . one of the neatest tropical towns that I have visited," its commerce largely based on the export of gold, rubber, indigo, ipecac and precious woods. In the early 1890s, several dredges were busy at Greytown with the excavation of an interoceanic canal, an attempt doomed by the American's pursuit of the Panama Canal.

Following almost a century of obscurity, Greytown's latest boom (literally), was its complete destruction during the recent Civil War, at the end of which New Greytown was established nearby with UN help. This is now a community of only about 300 people, as opposed to about 2,000 around 1910. Some of its citizens did not survive the war, many more relocated elsewhere, mostly in neighbouring Costa Rica, where they found safe haven and better economic opportunities. Ethnically, the new town now includes only a few creole people and native Indians, the bulk being people of mixed extraction. The community's economy is largely based on subsistence gardening, and the extraction of natural resources, notably fish, marine lobster and freshwater prawn, iguanas, and several species of vertebrate game, especially *javelinas* (Figure 2). Some coconut plantations in non-resident ownership along the coast, provide very limited employment.

By virtue of its proximity to the San Juan Delta Wildlife Refuge and control of the major access to the Rio Indio-Maiz Grand Biological Reserve, New Greytown, now finds itself again of strategic significance. In terms of area, biological wealth and natural integrity, these two protected areas represent both the geographic core and the crown jewels of Si-a-Paz. Together they cover about 3,500 square kilometres of pristine lowland tropical humid forest, the largest uninterrupted area of this type of ecosystem in Central America, in fact north of the Amazonian rainforest, and they abut the longest undeveloped beach in the Caribbean.

Two lifezones (tropical wet and premontane rainforests), and three transitional zones have been identified in this reserve, supplemented by an intricate system of natural barriers, canals, lagoons, as

FIGURE 2. In a subsistence community such as New Greytown, all economic activity involves the use of natural resources. The future welfare of this young man is intimately connected to the sustainability of local resources such as wildlife, both in its consumptive and non-consumptive aspects.

well as freshwater and brackish water swamps. This rich mix of riverine, estuarine, coastal lowland and hilly upland environments, results in conditions believed to have among the greatest biodiversity in the world (Anon. 1988, 1991).

This great natural reserve includes two completely pristine watersheds, and stable populations of many species of animals considered endangered in other regions, including five species of wildcats, the anteater, tapir, manatee, marine turtles, two species of crocodilians, freshwater shark and sawfish, the royal tarpon, macaws and harpyie eagle, just to name the more conspicuous. Its flora (estimated at 5,000 vascular plants), and invertebrate fauna (still largely unexplored), are certain to include species new to science.

As such, Greytown is predestined to become a hub for two new goldrushes, the one for biodiversity, the other focused on adventure tourists (ecotourists). National plans for tourism development envision historic Greytown, presently a sleeper, as a terminal along Nicaragua's southern tourist corridor, which stretches from Managua to Masaya, Granada, Ometepe Island, the Solentiname Archipelago, the Guatusos wetlands, San Carlos and El Castillo to Greytown, and from there connects with existing Costa Rican tourist bases at Barra del Colorado and Tortuguero. Thus, Greytown finds itself at the crucial junction of two axes, the East-West tourist corridor, and the North-South Paseo Pantera biodiversity corridor.

HISTORY OF PROJECT

As Si-a-Paz was starting to take shape in the late 1980s in conjunction with Central American peace initiatives, and with help from the Swedish, Norwegian and Dutch International Development Agencies, a needs assessment identified a biological field station in conjunction with the Indio-Maiz Reserve as one priority (Anon. 1988). The "Conceptual Framework and Plan of Action for the Development of Si-a-Paz," reaffirmed this need (Anon. 1991).

By that time a coalition consisting of Nicaragua's Ministry of Natural Resources (IRENA), The Wisconsin-Nicaragua Partners (of the Americas), and the University of Wisconsin at Stevens Point, formed, with the aim to help with development of this research station.

In 1991, coincident with Rotary International's then motto "Preserve Planet Earth," Rotary District 6220, representing 40 clubs in Wisconsin and Upper Michigan, and Rotary Managua, adopted this project as part of their World Community Service commitment.

Initial funds were made available for a feasibility study, and two groups of Rotarians ventured to Greytown in 1993 and 1994. The initial focus of these visits was to acquaint Rotary representatives with Nicaragua in general, and more specifically to generate understanding of and enthusiasm for the role of the station in the Si-a-Paz scheme in sustainable development of this region. The inclusion of a civil engineer, environmental architect, an ecotourist agent, as well as marketing, fisheries and resource management professionals, and several generalists, was to assure representation of relevant and diverse points of view.

Numerous meetings involving governmental officials, and a number of local NGOs, most notably the Association for the Development of Greytown (APDG), the Council of Evangelical Churches of Nicaragua (CEPAD), and the Fundacion del Rio, proved extremely valuable in assessing the general situation and feasibility of the project.

In the course of the two visits, several potential sites for the station were explored along the lower Rio San Juan, and near Dos Bocas de Rio Indio. Although physically suitable, these sites were, after lengthy discussions, rejected in favor of a location just north of New Greytown on the west shore of the Rio Indio. By developing the station near town, but with easy access to the core of the protected areas, it will be in an excellent position to not only allow basic research, such as biological inventories and ecological studies in one of the world's most complex ecosystems, but at the same time applied research for the solution of immediate community needs, some of them spelled out by Ubau et al. (1993), or communicated by Greytown community leaders.

The opportunistic extraction of fish, prawn, lobster, game, and various wild plant products, as presently practised in the surroundings of Greytown, seems to be sustainable at present rates of use. A growing Greytown, however, envisioned to eventually reach 1,000 people, is likely to put greater strain on the fragile natural support systems of the neighbouring protected areas, as animal and plant resources near the community diminish through increasing rates of harvesting. As a result, researching these resources and associated social and economic patterns of the user population, quite aside from providing employment for station staff, guides and suppliers of food and other services, will provide a sound scientific basis for the development of sustainiable management plans. Other needs and opportunities for applied research pertain to the rehabilitation of local coconut plantations, exploratory research into crops such as ginger or cocoa, and possibly captive wildlife (butterflies, iguanas and tropical birds) for local use or sanctioned trade. Ideally, as these various ventures start bearing fruit, and ecotourism (Talavera and Cuevas 1994) of the non-consumptive and consumptive (sport fishing and hunting) varieties provides income, resource extraction will increasingly be regulated, as determined by local participation in planning and management, and with advice from scientific authorities. All these activities are likely to contribute to better village infrastructure, and with higher levels of health care, living standards, education, and ultimately restrained growth rates in the human population. The overall effect of such community development will hopefully be sustainable land use coupled with optimum protection of the wildlands.

Besides data and direct or indirect economic benefits, the field station will also contribute to environmental education of the community, with the intent to strengthen local resolve in safeguarding the integrity of *their* natural systems, and to enhance the effectiveness of resident naturalists. Increased traffic and communication with the outside world, for better or worse, is another likely effect resulting from the establishment of a station in what is still a very difficult to reach, remote community.

STATUS

At this stage, the feasibility of the project was endorsed by all participating parties. The community of New Greytown committed four hectares of land, and promised labour and some materials for its construction. A cost estimate and basic architectural plans were already drawn up by Ubau et al. (1993), but there remains considerable room for consolidation and elaboration of these plans, and questions of funding, timeframes, responsibilities, and eventual ownership of the station, need to be clarified.

To have reached this stage has already taken an inordinate amount of time and effort, largely based on difficulties inherent in the remoteness of Greytown, as well as Nicaragua's difficult post-war society and economy. Among these, the most vexing were unreliable communication and lack of coherence of parties concerned, especially for lack of continuity in government coordinators. To date, three ministers of natural resources, and an equal number of Si-a-Paz coordinators have been involved with various aspects of the project, but never long enough to carry through. As a result, NGOs increasingly assumed greater roles, fed by a growing sense of responsibility and independence, budding entrepreneurial spirit, and better cohesion among those involved in planning regional and local projects.

PROSPECTS

The most likely scenario for a Greytown research station, is for an at least initially modest, but solid (with respect to termites, weather and sea air corrosiveness) facility, able to comfortably accommodate the needs of pioneer scientists. Evolving priorities will then dictate any modular expansions considered necessary. In this, the range of possibilities is considerable, as reflected in existing research stations elsewhere. The "Guide to Biological Field-stations" (Merritt and Hannakan 1992), for instance includes 16 stations in Central America and the Caribbean, some with a very narrow focus, others more general, some very rudimentary, others, such as La Selva in Costa Rica, at the cutting edge with the latest in data processing, library support, GIS technology, and a staff of 30. None of the stations listed, however, serves the needs of an individual community in the way the Greytown station is envisioned.

The trend for greater regional and local control, in combination with better political and economic stability that may result from next year's national elections, suggest that conditions for implementing the Greytown station and other projects may become more favorable. As before, their ultimate success will, however, depend to a considerable degree on Nicaraguan initiative, cohesion and participation under the supervision of a local coordinator. Funding will then, at least in part, come from cooperators abroad.

REFERENCES

Anon. 1988. Propuesta para la creacion del Sistema Internacional de Areas Protegidas Para La Paz (Si-a-Paz) en el area del Rio San Juan. Comisiones Nacionales de Nicaragua y Costa Rica Sobre Reservas Fronterizas Para La Paz. 58 p. (mimeo).

Anon. 1991. Conceptual framework and plan of action for the development of the International System of Protected Areas For Peace (Si-a-Paz). IRENA, Managua, Nicaragua 201 p. (mimeo).

Anon. 1993. Plan de accion forestal de Nicaragua (PAF-NIC), and Plan de accion ambiental de Nicaragua (PAA-NIC). IRENA, Managua, Nicaragua (mimeo).

Belt T. 1874. The naturalist in Nicaragua. Edward Bumpus. London, UK.

Brannstrom, C. 1992. Almost a canal: visions for interoceanic communication across southern Nicaragua. MSc Thesis, Geography Dept., University of Wisconsin, Madison.

Girot, P. O. And B. Q. Nietschmann. 1992. The Rio San Juan. Nat. Geogr. Res. Expl. 8: 52-3.

Marynowski, S. 1992. Paseo Pantera, Wild Earth (Special Issue on Wildlands): 71-74.

Merritt J. F. and C. J. Hannakan (eds.). 1992. Guide to biological field stations. Carnegie Museum of Natural History, Rector, PA.

Talavera, V. C. and V. C. Cuevas. 1994. Proposal of community based sustainable tourism municipality of San Juan del Norte. San Juan River, Nicaragua 7 p. (mimeo).

Ubau, L., R. Solo and Z. Acevedo. 1993. Acciones para facilitar el desarrollo socioeconomico de San Juan del Norte. Planas pare el manejo y proteccion de las areas de reserva del municipio. IRENA, Managua, Nicaragua 43 p. (mimeo).

Wille, C. 1991. Peace is hell. Audubon, 93: 63-70.

World Commission on Environment and Development. 1987. Our common future. Oxford University Press, Oxford, UK.

Sustainable Forests:
It's About Time (Montana)

Barry R. Flamm

ABSTRACT. Forest health should be determined by ecological criteria as opposed to the more limited tree production approach. We must recognize the vital relationships between conserving biological diversity and sustaining forest ecosystems. World-wide forest management practices have too often ignored ecological principles, thereby jeopardizing forest health in the long-term. Much warranted attention has been given to rain forest problems. Temperate, mountain forests are also threatened, presenting unique sustainability problems. The forests of western Montana are a case in point. Sustainability is, of course, about time, and it is about time that forest management is changed to assure healthy forests for the future. *[Article copies available for a fee from The Haworth Document Delivery Service: 1-800-342-9678. E-mail address: getinfo@haworth.com]*

INTRODUCTION

Throughout the world, forests have been degraded through over-exploitation, poor forestry practices and pollution. As if this were not enough, in recent years we have been haunted by the spectre of global climate change with predicted massive impacts on forests

Barry R. Flamm is Natural Resources and Environmental Consultant, Montana, USA.

[Haworth co-indexing entry note]: "Sustainable Forests: It's About Time (Montana)." Flamm, Barry R. Co-published simultaneously in *Journal of Sustainable Forestry* (Food Products Press, an imprint of The Haworth Press, Inc.) Vol. 4, No. 3/4, 1997, pp. 139-147; and: *Sustainable Forests: Global Challenges and Local Solutions* (ed: O. Thomas Bouman, and David G. Brand) Food Products Press, an imprint of The Haworth Press, Inc., 1997, pp. 139-147. Single or multiple copies of this article are available for a fee from The Haworth Document Delivery Service [1-800-342-9678, 9:00 a.m. - 5:00 p.m. (EST). E-mail address: getinfo@haworth.com].

caused in part by the very loss and degradation of forests. As we here know, when we destroy a forest we destroy not only its trees but its ability to provide food, fuel, important products for commerce, biological diversity, and critical ecological services such as watershed protection and climate regulation.

Sustainable forests is the imperative for the future and the greatest challenge for foresters. We must ensure sustainability of the entire forest, necessary if we are to have sustainable yields of all important forest resources. We must have sustainable yields to maintain sustainable economies and sustainable societies (Maser 1988).

Much warranted attention has been given to tropical forests' destruction. Temperate, mountain forests are also threatened, presenting unique sustainability problems. For example, 65% of the forests of Nepal have been cleared or degraded resulting in severe watershed, soil erosion, and supply problems.

In today's remarks, I will concentrate on sustainability issues on forests closer to this conference site–the forests of Montana, USA, and in particular the Flathead National Forest.

The title of this paper has an obvious double meaning; it reveals my concern for the future and my impatience with my own profession. Sustainability is of course about time and it is about time that forest management is changed to assure better forest ecosystems for the future.

MONTANA

Montana is considered by many Americans to still be a wild frontier state. Indeed, Montanans themselves glory in this image. Although it's US's fourth largest state in area (147,046 sq. mi.), it has only 808,487 people (5.55 people per sq. mile) (World Almanac 1993). Yet parts of the state are growing rapidly. In particular, the Flathead National Forest is located in a rapidly growing area of northwest Montana resulting in land use and social stresses.

Western Montana is noted for its scenic beauty, clear waters and air and the availability of a variety of exceptional outdoor recreational opportunities–a Mecca for tourists. These values often conflict with natural resource development.

I believe it is fair to say that as a whole, Montanans over the last century were preoccupied and dominated by commodity extraction interests: timber, mining and grazing. Contemporary politicians are still fond of saying, "Montana is a natural resource state–we must develop our resources." Today this is an economic myth made obsolete by the real facts of the Montana economy as it has grown and evolved. In 1989 mining and wood products contributed only 4% each to total private earnings (Darrow 1995).

FLATHEAD NATIONAL FOREST

The 920,000 hectares Flathead National Forest is in northwestern Montana. It is predominantly forested–*Pseudotsuga menziesii, Pinus ponderosa, Larix occidentalis, Abies grandis, Pinus monticola, Pinus contorta, Thuja plicata, Abies lasiocarpa,* and *Picea engelmannii. Pinus albicaulis* and *Larix lyallii* occur at high elevations. The suitable land base for commercial timber is 268,000 hectares. Approximately half of the forest is designated wilderness. The remaining land area that is designated unsuitable is high country, grizzly bear protection areas, recreation areas and administrative sites (Flathead National Forest 1992).

The utilization of timber resources on the Flathead National Forest began in the early 1900s with the establishment of a sawmill on Flathead Lake at Somers. In 1913 the largest sale in the history of the forest was made in the vicinity of Swan Lake Guard Station–a total of 91 million board feet of ponderosa pine and larch was harvested (Flathead Nation Forest 1974). After this surge, timber activity was at a low ebb until the beginning of WWII. During this period selection system of silviculture was primarily used. Post-WWII timber activity picked up to meet housing shortages. In 1953 several large drainages were infected with spruce bark beetles and large scale salvage was initiated. These stands were primarily clearcut–a practice that continues into the 1990s.

Timber planners are an optimistic lot. In the 1969 timber management plan they established an annual harvest of 194 million board feet–up from 137 million board feet in the previous plan (Flathead National Forest 1974). On January 1, 1970, the National Environment Policy Act was signed into law requiring among other things, an Environmental Impact Statement (EIS) and opening up decision-making to the public. Timber cutting would be more carefully watched for environmental effects.

In 1976 another important piece of legislation passed: the National Forest Management Act, which required each forest to develop a comprehensive plan involving all uses. It took until January 22, 1986 for the Flathead National Forest Plan to be approved. It called for a harvest of 100 million board feet (Figure 1). The plan was appealed contending that inadequate habitat would be maintained for management indicator species (MIS) and for viable wildlife populations. Finally a lawsuit against the forest plan resulted in a court order to adjust the allowable sale quantity (ASQ) of timber in a manner to ensure conservation of endangered species. In 1995, the forest supervisor amended the forest plan reducing the annual planned cut to 54 million board feet (Flathead National Forest 1995).

The timber volume actually sold and harvested differs from the planned sale quantities, dropping precipitously since 1988 (Figure 2). As would be expected, acres harvested have also declined (Figure 3). Planned silviculture method (1995-99) remains predominantly even age (18,456 acres). Uneven age planned is 530 acres (Figure 4).

The accumulated effects from the years of heavy cutting resulted

FIGURE 1. Planned sale quantities Flathead National Forest

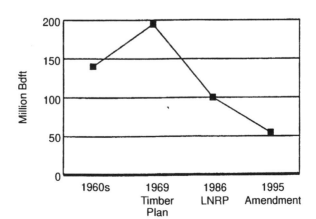

FIGURE 2. Timber volume sold and harvested Flathead National Forest

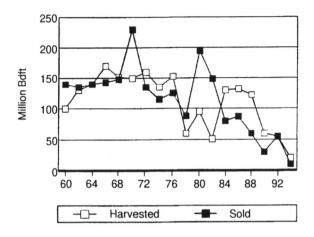

FIGURE 3. Harvest history in acres Flathead National Forest

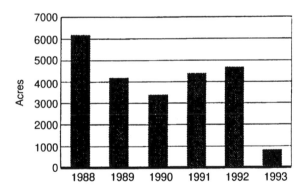

in a distinctly different landscape–clearcutting was the predominant silvicultural system used. Most of the suitable forest land base was heavily roaded, impacting fish and wildlife resources. The planned 1995-1999 road densities differ little from the previous five years (Figure 5).

Grizzly bears are adversely affected when open road density exceeds one mile per square mile. The amended plan complies with

FIGURE 4. Planned harvest acres, 1995-99, by method Flathead National Forest

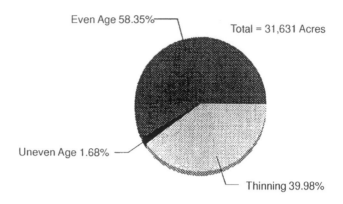

FIGURE 5. Miles of road Flathead National Forest

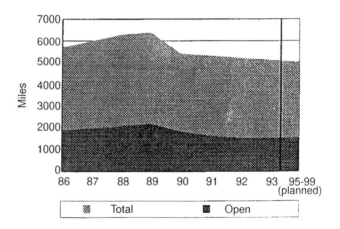

the Fish and Wildlife Service's biological opinion of 1/6/95 for grizzly bear recovery. Under this plan there will be no net loss or net gain in motorized density. The area of high density (>1 mile/sq. mile) open motorized access will be limited to no more than 19% of the MS-1 and MS-2 within a bear management unit. (Note: MS-1 = Management situation 1 are areas key to the survival and recovery of the grizzly bear population. MS-2 = Management situation 2 are areas which lack distinct grizzly bear population centers. Habitat either not needed or not yet determined as needed for grizzly bear survival and recovery.) This should also benefit other wildlife, as well as fish and water quality.

Throughout Montana stream ecosystem degradation is severe and is continuing almost unabated. Consequently both fish and water quality and quantity are in a growing state of crisis. Unique fish populations which were healthy for thousands of years are today in serious trouble; virtually all native fish in the Northern Rockies are imperiled (Figure 6). Stream ecosystem degradation is the most controversial problem: loss of riparian area, loss of large pools, sedimentation and pollution, and dewatering. Logging is one of the causes of this degradation.

US Forest Service research has established that there is no "safe" level of sedimentation that does not affect bull trout spawning (US Fish and Wildlife Service has determined that bull trout warrants protection under the Endangered Species Act). Yet the Flathead plan still allows increases in fine sediment levels that can degrade spawning and rearing habitat in the highest quality streams (Pacific Rivers Council Inc. 1995). Past over-cutting and poor practices have adversely affected the forest ecosystem, degraded scenic beauty and affected the socioeconomic structure of communities.

FIGURE 6. Native fish at risk in Montana Flathead National Forest

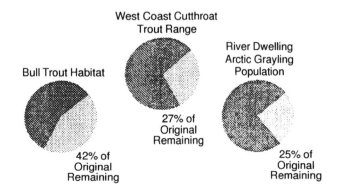

West Coast Cutthroat Trout Range

River Dwelling Arctic Grayling Population

Bull Trout Habitat

27% of Original Remaining

42% of Original Remaining

25% of Original Remaining

CONCLUSION

The Forest Plan amendment is a giant step forward. The reduction of annual sale quantity (ASQ) to 54 million board feet provides better opportunity for managers to protect other forest resources. It has the effect of increasing average rotation age of the forest plan from 106 year rotation to 200 to 400 year rotation, allowing for a more diverse forest structure, recovery time and a safety net for unforeseen natural and human events. It encourages modification of current clearcutting practices. In all but lodgepole type forests—Flathead National Forest silviculturalists are prescribing partial cuts. These are essentially seed tree or shelter wood cuts with the over story left to provide habitat diversity, stand structure diversity, and to provide a source of dead standing and down wood vital to proper ecosystem functioning.

Less habitat fragmentation will thus occur. Areas will be restored and roads closed and rehabilitated. This all is a big plus. However, does it go far enough?

The wildlife, fish, recreation, scenic, and water quality values of the forest are extremely important. The greater Flathead-Waterton-Glacier ecosystem is one of the world's jewels. The substantial reduction in harvests may still not be sufficient due to the cumulative effects of past actions. In other words, the reduction may have come too late. Timber planners are having trouble finding areas to cut and getting the sales past an increasingly vigilant public.

An example of the difficulty of finding sale areas is exemplified by the Crane Mountain sale adjacent to the very beautiful and popular Flathead Lake. The scenic effects are large plus possible lake water degradation.

The sale comes down to private land which lies between the forest and lake shore. Many people have expressed concerns over this sale. An industry solution to public appeals is through the US political process—to change environmental laws. One bill would exempt logging from all environmental laws including NEPA and the Endangered Species Act and would eliminate administrative and judicial review of timber sales and suspend the rights of citizens to challenge government decisions—a serious step backwards for US conservation efforts.

FUTURE OBJECTIVES

Our forest management objective must be to ensure self-sustaining and self-repairing forests (Maser 1994). Protection of ecosystem integrity is a fundamental principle. We must keep all the parts.

The goal should be non-degradation of the ecosystem. In addition, large areas in all forest types must be protected from development to provide a reservoir of biological diversity. These areas should be linked by corridors where management practices have been modified to allow biological movement and species protection. The approach: work with nature, not try to conquer her.

REFERENCES

Darrow, G. 1995. Montana–A State of Entrepreneurs. Unpublished.

Flathead Forest. 1974. Timber Management Plan, Kalispell, MT, USA.

_____ 1992. Land and Resource Management Plan: Supplemental Monitoring Report, Kalispell, MT, USA.

_____ 1995. Forest Plan Amendment #19. Kalispell, MT, USA.

Maser, C. 1988. The Redesigned Forest. R&E Miles, San Pedro, CA, USA.

_____ 1994. Sustainable Forestry: Philosophy, Science and Economics. St. Lucie Press, Delray Beach, FL, USA.

Pacific Rivers Council Inc. 1995. Northern Rockies Forests and Endangered Native Fish. Eugene, OR, USA.

World Almanac (The). 1993. Pharos Books, New York, NY, USA.

Partnership Building
for Sustainable Development:
A First Nations Perspective from Ontario

P. A. Story
F. H. Lickers

ABSTRACT. First Nations' approaches toward environmental stewardship have always been based upon partnership and a sense of belonging within the natural environment. The cornerstone of traditional community relationships is the three-pronged model of partnership building, known to the Iroquois nations as the "Zeal to Deal." The Eastern Ontario Model Forest (EOMF) was initiated as one of ten model forest sites across Canada, under a federal sustainable development initiative. The Eastern Ontario Model Forest was formed from people with many different viewpoints, ideals, tools, and ideas. A mixture of people holding local landowner values, native philosophies, and scientific knowledge were brought together to develop a working partnership at a community level. This presentation will analyze approaches in facilitating a partnership according to traditional knowledge of the Iroquois nations (University of Ottawa, 1994). This presentation will show an analysis of the working relationships of people and organizations according to tools developed and practiced by Aboriginal peoples of Eastern Ontario. These tools include the inclusion of each partner in the Eastern Ontario

P. A. Story is Project Coordinator for the Eastern Ontario Model Forest, Ontario, Canada.

F. H. Lickers is Director, Department of the Environment, Mohawk Council of Akwesasne and Director, Eastern Ontario Forest Group, Inc., Ontario, Canada.

[Haworth co-indexing entry note]: "Partnership Building for Sustainable Development: A First Nations Perspective from Ontario." Story, P. A. and F. H. Lickers. Co-published simultaneously in *Journal of Sustainable Forestry* (Food Products Press, an imprint of The Haworth Press, Inc.) Vol. 4, No. 3/4, 1997, pp. 149-162; and: *Sustainable Forests: Global Challenges and Local Solutions* (ed: O. Thomas Bouman, and David G. Brand) Food Products Press, an imprint of The Haworth Press, Inc., 1997, pp. 149-162. Single or multiple copies of this article are available for a fee from The Haworth Document Delivery Service [1-800-342-9678, 9:00 a.m. - 5:00 p.m. (EST). E-mail address: getinfo@haworth.com].

149

Model Forest as a full partnership based on the "Zeal to Deal." The development of partnerships from the use of all three tools in this community-level organization will be analyzed. Challenges in the development of new partnerships, and limitations of partnerships according to naturalized knowledge systems will be analyzed. The strength of the EOMF organization based on its diversity will also be addressed. *[Article copies available for a fee from The Haworth Document Delivery Service: 1-800-342-9678. E-mail address: getinfo@haworth.com]*

THE EASTERN ONTARIO MODEL FOREST: LEARNING THE TOOLS

The Eastern Ontario Model Forest is one of ten working examples of "sustainable forestry" in Canada. The Canadian Model Forest Program was developed in 1992 by the Canadian Forest Service of the Department of Natural Resources (then called Forestry Canada). The Eastern Ontario Model Forest incorporates 1.5 million ha of public, private, municipal and aboriginal lands in the counties of Prescott, Russell, Stormont, Dundas, Glengarry, Lanark, Leeds, Grenville, Regional Municipality of Ottawa-Carleton, and the Mohawk Community of Akwesasne.

The direction of the Eastern Ontario Model Forest is by an elected nine-member board of directors. Three directors are appointed, six are elected by a membership body. The three appointed positions are from the three "key" stakeholders: The Ontario Ministry of Natural Resources (Government Rep.), Domtar Specialty Fine Papers (Industry Rep.) and the Mohawk Council of Akwesasne (First Nations Rep.). At all times, at least one representative at the Board of Directors level must be a member of the aboriginal community. This structure has allowed the First Nations policies, concepts and teachings to be a key part of the decision making body of the Eastern Ontario Model Forest.

One of the key projects in the Eastern Ontario Model Forest has been the "Akwesasne Partnership." This project has gone far beyond a traditional project with goals, objectives, tasks, results, and reports. One of this project's key mandates is to incorporate traditional knowledge into the decision making, policy making, planning, evaluation, and operating mechanisms of the program. The partnership is growing, and the lessons taught to the non-aboriginal partners of the EOMF program are cornerstones to the operation of the EOMF as a whole.

Two key concepts and two key processes of aboriginal traditional knowledge or "naturalized knowledge systems" have been examined here in greater detail. These concepts and processes have been introduced to the EOMF organization and have been used in various places for the organization, planning, decision making, and technology transfer/education phases of the program.

The processes investigated further are:

- the *zeal to deal*: the process of solidifying working relationships between partners, and potential partners;
- decision making by *consensus* or mutual agreement, rather than a vote system.

The *concepts* investigated further are:

- *time*: Planning for the "seventh generation"
- *scale*: naturalized knowledge systems (aboriginal theory); matching the "scale of government" to the scale of the land area being managed.

The analysis of the EOMF using the concepts and processes is indicated by "successes" and "challenges." These concepts and processes have been used both successfully and unsuccessfully in decision-making processes, partnership building, planning, and implementation of projects.

The analysis is at the end of the explanation of each process or concept. Successes are marked with an "S" and challenges are marked with a "C."

PROCESS NUMBER ONE: THE ZEAL TO DEAL

This process is *the* most powerful concept in the aboriginal traditional knowledge taught to the EOMF. Basic aboriginal thought promotes the ideal that "cooperation is the only way to survive." Individuals must be committed partners in the cooperation of the care for our natural resources. The ideology of the "zeal to deal" is integral in solidifying relationships between potential partners.

There are three elements to this process: respect, equity and empowerment. These elements must be used in the proper order and in the proper proportions for a successful partnership process.

Respect

The process of working with First Nations communities requires that both cultures (aboriginal and non-aboriginal) must develop a basic respect or understanding of each other. Time must be taken for non-aboriginals to meet, learn and develop a friendship and working relationship with aboriginal community leaders. Concepts of environment, law, government, and society all differ radically from mainstream north American culture. These differences *must* be acknowledged, understood and respected. Only through the sharing of knowledge and experience can the aboriginal and non-aboriginal groups work together and successfully complete a project. Respect also includes the "principle" of inclusion, where all new projects and partners are given equal opportunity to present their ideas, and be heard.

Equity

After the initial introduction between cultures, "equity" will become an important next step. Equity can be defined as "anything that has value." Therefore, equity may be money, labour, time, materials, interest, support, etc. In western societies, equity usually indicates financial equity, or, simple dollars. It is important that we begin to value information, knowledge, and "sweat equity" (donated time) as equivalent contributions to a partnership project, as some community based, volunteer, or aboriginal organizations may not have true "dollars" upon which to build a partnership.

Empowerment

Empowerment has been defined as "the act of enabling" (University of Ottawa, 1994). This means, that any partnership with a "host" organization requires that the "host" organization must allow the partner to undertake and complete the project on its own terms, and with its own particular style. This is the hardest of all of the three concepts to implement, as it requires trust between potential partners. Without respect, and equity, trust cannot be built.

Putting Together the Deal

The three elements explained above must be put together in appropriate order and appropriate proportions. It is important to have all three elements involved in the process.

Figure 1 indicates the three elements working together, and the appropriate order, and proportion of the three elements to best generate the deal.

A Deal That Is Not Working

There are situations that can arise if only some of the elements are in place. A partnership built on respect only, will paralyze the potential partner, and maintain a type of "patronizing" approach. A partnership built without equity, will limit the potential partner's feelings about him/herself as an equal partner. It will also limit how much the partner is able to provide input and guidance to the partnership. A deal without empowerment will only maintain a "control" relationship, with the partner feeling disenfranchised from the decision-making and overall direction of the project. A deal without trust is also doomed to fail, as there will be a basic misunderstand-

FIGURE 1. Elements in the "Zeal to Deal," their appropriate order, and appropriate proportions for a successful deal.

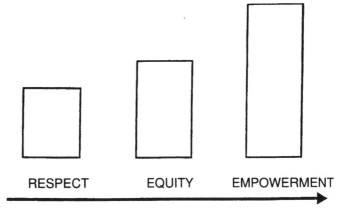

Source: University of Ottawa, 1994

ing of the potential cooperators, their interests, backgrounds, working styles, key issues, interests, and concerns. Disagreements, arguments, and other unpleasant working relationships often arise in attempted "deals" without respect.

Maintaining the Deal

Over time, the deal must be re-introduced, and restrengthened. Time is required to make sure respect for the project, the partnership, and the partners involved continues.

Evidence the Deal Is Working

There are various signs that the "deal" is in place. Some of the most obvious indicators are: generation of new equity over time, increase in number of participants, a feeling of satisfaction and contentment with the progress of the project, increase in output, and general increase in interest in the partnership.

ANALYSIS OF THE EOMF USING THIS PROCESS

Increase in Equity

S   The EOMF has been successful in levering additional projects to increase the overall scope of the program. Four additional Green Plan projects, the Calakmul Model Forest twinning, and "Jobs Ontario" funding have elevated the financial equity of the EOMF partners and program.

S   EOMF partners have also been able to lever equity by their involvement with the EOMF program.

S   There is also an increasing interest in volunteering with the program. To date, we have three volunteers who have had an interest in helping with special events and projects.

Increase in Cooperation

S   Fifty projects have been initiated and are underway in the EOMF. Some have already been completed, or merged with others in the development of the EOMF.

S   Thirty-six projects are underway as of April 1995.

Inclusion

S   All new projects and partners are welcomed to the table, and are encouraged to participate.

C   Finding time for new projects, and financial limitations are sometimes a hindrance to allowing fair access to the program for all new ideas.

S   Evaluation framework for the EOMF was developed with the help of a cross-section of partners, supporters, members and interested people. A two day seminar directed by a professional evaluator was hosted for these groups. A framework was designed from the compilation of ideas and opinions.

Maintaining the Deal

C   Time is a limiting factor in maintaining the deal. There is not enough time to sometimes adequately maintain the deals already established. Some dissatisfaction sometimes occurs amongst the partners.

C   Three projects have been discontinued due to a lack of respect and/or equity.

Project Organization

S   The EOMF projects can be divided into these three elements of "respect," "equity" and "empowerment." Figure 2 divides these projects and shows how they interact. This process is interlinking, and feeds back upon itself to strengthen the deal, and hence strengthen the EOMF program as a whole.

S   One committee, the EOMF Trails and Outdoor Education Committee, has been developed into a conceptual diagram (Figure 3) to show in detail how this process works. This committee is a working example of a successful "deal."

FIGURE 2. Sample projects/associated activites as divided into the three elements. These projects feed into each other, and back into the program.

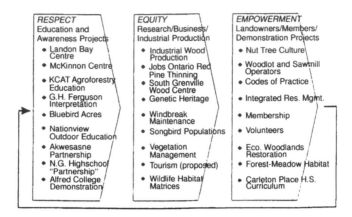

FIGURE 3. Trails and Outdoor Education Committee example.

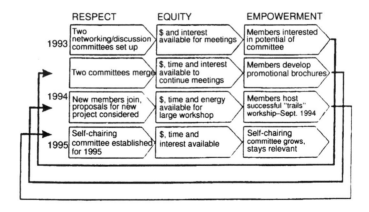

PROCESS NUMBER TWO:
CONSENSUS DECISION MAKING

Consensus

Consensus, defined as "a decision making process where unanimous support is required," is a process long-used by North American aboriginal communities. In these societies, the traditional western confrontational style of decision making is considered inappropriate, and offensive. In the mainstream "traditional" confrontational style of decision-making, consensus style decision making is sometimes seen as "revolutionary" (Kerr-Wilson 1995).

The process is time consuming, yet not tremendously complex. Each member at the decision-making table is allowed or encouraged to speak his or her mind, and form an opinion on the topic up for discussion. Rationale behind the decision or the opinion is often sought from the other members. As opinions and ideas are put forward, often there is some dissent between two or more members. Further discussions into the rationale, background, thoughts, feelings, and experiences of the dissenting persons are encouraged and explored. It is the responsibility of the group to encourage the information sharing, and to support the discussion where necessary.

Consultation with Others

Dissent is sometimes inevitable. In these cases, the issue is often brought back to the overall driving issue of the gathering. A reminder of the goal of the organization, and the decision making body is sometimes brought forward as a focus to steer the opinions and discussions.

Freedom of Information

The inclusion of a wide variety of stakeholders, partners, and individuals in the decision making process must be followed by the free dissemination of information learned from the process. The Eastern Ontario Model Forest has adopted a concept from aboriginal traditional thought: "knowledge has power only if it is shared."

ANALYSIS OF THE EOMF USING THIS PROCESS

Use of Consensus in Decision Making

C   Initial EOMF Board of Directors' meetings involved voting, confrontation, and dissent.

S   Board of Directors' meetings now operate on a consensus style of operation. (Board meetings can become very lengthy however!)

S   Some project management sub-committees are now using this approach.

C   Occasionally, dissent occurs, and some decisions are not always a true "consensus."

Consultation with Partners and Peers

S   Project management committees (subcommittees of partners, supporters, EOMF staff, board, and Canadian Forest Service representatives) are frequently consulted regarding information and issues on project management. Sometimes decisions are deferred until more information from those affected is gathered.

S   Informal committees for the management of Forestry Standards, Ecological Woodlands, Industrial Wood Production, Relative Density Guidelines, Volunteers, and Membership.

S   Formal committees are set up to deal with Forest Science Review, Finance and operations, Public Information and Education, Board of Directors Nominations, and Trails and Outdoor Education.

Freedom of Information

S   The current policy for information dispersal is to promote, disseminate, and share information discovered in each of the EOMF projects.

C   The western concept of "ownership of information" sometimes clashes with traditional aboriginal thought. Anything developed as an EOMF product is freely disseminated. Projects with cooperative funds, may cause some tension over the ownership and dissemination of project results.

CONCEPT NUMBER ONE: TIME

Traditional aboriginal thought includes the use of a time frame that is much longer than much of the "western" thought or planning scales. Aboriginal communities, have developed and lived by the phrase "planning for seven generations," which has come to be used more and more often in sustainable development of our resources, our world, and now into sustainable forestry.

The First Nation decision-making process requires a time frame upon which to focus. Consensus is the method of decision-making that is used when deciding upon a course of action, however, this is not the only consideration when making a decision. Acknowledgement of a time frame for decision-making is just as important as acknowledging the reasoning behind a particular choice. Aboriginal societies, particularly the Iroquois nations of eastern North America, have taken into consideration the future seven generations of humans that will inhabit the Earth long after the existing generations have passed on. Each successive generation is considered more important than the next, thereby a sustainable future for the yet unborn can be assured. This seven generation planning framework mimics natural processes such as the maturation of a forest, or the life of a tree. The idea of seven generation planning allows humans, who have a difficult time imagining a time frame beyond one's own existence, to begin to mentally grasp time frames of natural processes.

Today's society is built upon profits, short-term survival, and competition. In short-term planning frameworks, environmental destruction can be justified. For short-term financial sutainability, "pillaging" our forests, and quickly liquidating them into capital, ensures market sustainability, and operating profits for a corporation. As we all know, short-term thinking has caused much of the environmental destruction around us. Adoption of a long-term, seven generations planning framework is often foreign to many corporations, individuals, organizations, and political parties in today's western society.

ANALYSIS OF THE EOMF USING THIS CONCEPT

S   EOMF has adopted the motto "a forest for seven generations" to help guide the direction, leadership, and method of
thinking regarding project progress, and project approval.

S   The EOMF has initiated a long-term strategic planning committee to identify long-term potential for the group, a long-
term strategy, and funding considerations to extend the hard
work of the EOMF to date to an indefinite time into the
future.

C   Funding, reporting and project management runs on a
12-month, fiscal year basis. Projects require visible results,
and must have a cleared budget every 12 months. The entire
Model Forest program currently has a five-year mandate.
Development of "sustainable" forestry efforts requires a
much longer time frame. This generates a lot of pressure for
groups to come together and produce valid results.

C   Current funding of the program is heavily reliant on public
funding. Most government initiatives operate on a four-year
mandate. As much environmental programming, operations, and research is government funded, research and operational activities and personnel for forestry programs
often change with a new government party. This makes it
difficult to truly plan for that seventh generation.

CONCEPT NUMBER TWO: SCALE

A growing body of research in both anthropological and environmental circles is the development and promotion of what has been
termed "naturalized knowledge systems."

"Naturalized knowledge systems" refer to the conceptual paradigm of hierarchies, social structures, levels of knowledge, and
governance from indigenous societies. The word "naturalized" refers to the natural, original inhabitants of an area, who hold a wealth
of knowledge about their traditional homelands. A vast amount of
this information, that many consulting firms, research organizations, government programs, and university level studies are seeking, may be known or enhanced by traditional knowledge of indige-

nous peoples. Much of this information is in a verbal, or oral tradition, and as such, cannot be researched through traditional European written histories.

One of the central ideas in this naturalized knowledge system is the idea of "scale." An organization to manage environmental systems, for example, would attempt to manage at a level appropriate to the ecosystem unit. The level of "community" is appropriate for this type of management. Other levels of concern in the aboriginal naturalized knowledge system include the subcellular, cellular, individual, group, nation, confederacy and spiritual realm. Each of these has special areas of emphasis for governance and management. Figure 4 compares these scale levels.

The Eastern Ontario Model Forest has been referred to as a "grass-roots" community forestry program. The entire Canadian and International Model Forest program is attempting to manage forests for a variety of competing interests for a long-term. The scale unit of community level management is appropriate here.

ANALYSIS OF THE EOMF USING THIS CONCEPT

S   EOMF uses the community level approach to forest management. Important players in the EOMF are the private

FIGURE 4. Scales in the Naturalized Knowledge System.

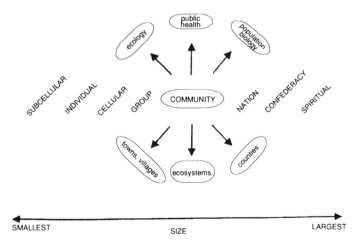

landowners, and associations that support private landowners in the local area.

C   The EOMF program sometimes has conflicting areas of influence with some Natural Resource organizations that manage on a national scale. Mandates sometimes conflict with the Provincial Ministry of Natural Resources.

S   Community groups fit well in the EOMF concept. Some of the most powerful results in EOMF projects are from cooperative community groups.

S   Like an ecosystem, the EOMF strives to be diverse in its approach to forestry. The more diverse an ecosystem, the more healthy the ecosystem. The same approach has been taken with the strength of the EOMF program for managing multiple values.

REFERENCES

University of Ottawa, 1994. First Nations–Environmental Knowledge and Approaches to Natural Resources.

Kerr-Wilson, John. 1995. Personal Communications.

Partnership Building
for Sustainable Development:
An Industry Perspective
from Saskatchewan

Jack Spencer

ABSTRACT. Weyerhaeuser chose to become a partner of the Prince Albert Model Forest primarily for two reasons: to strengthen relationships with other forest stakeholders, most notably the Montreal Lake Cree Nation, and to create in the Model Forest an accurate database about Saskatchewan forests.

Both of these goals have been and continue to be realized. Montreal Lake and other forest users have participated in projects with Weyerhaeuser. The company provides technical training and jobs to First Nations people. Weyerhaeuser has also begun to incorporate some research results into its operations as a means of addressing environmental impacts on biodiversity and wildlife habitat. *[Article copies available for a fee from The Haworth Document Delivery Service: 1-800-342-9678. E-mail address: getinfo@haworth.com]*

HISTORICAL CONTEXT

Weyerhaeuser's involvement in the Prince Albert Model Forest began in 1990 with the announcement of the Green Plan and refer-

Jack Spencer is Chief Forester with Weyerhaeuser Canada Ltd's. Saskatchewan Division, Saskatchewan, Canada.

[Haworth co-indexing entry note]: "Partnership Building for Sustainable Development: An Industry Perspective from Saskatchewan." Spencer, Jack. Co-published simultaneously in *Journal of Sustainable Forestry* (Food Products Press, an imprint of The Haworth Press, Inc.) Vol. 4, No. 3/4, 1997, pp. 163-169; and: *Sustainable Forests: Global Challenges and Local Solutions* (ed: O. Thomas Bouman, and David G. Brand) Food Products Press, an imprint of The Haworth Press, Inc., 1997, pp. 163-169. Single or multiple copies of this article are available for a fee from The Haworth Document Delivery Service [1-800-342-9678, 9:00 a.m. - 5:00 p.m. (EST). E-mail address: getinfo@haworth.com].

163

ence to a program called "Model Forests." In March of 1991 we sent a letter of interest to the minister about this program. When Forestry Canada came to Prince Albert in September of 1991, however, I have to admit that we were somewhat reticent. Although we recognized the potential of the program, we also recognized the significant effort involved and the commitment required just to prepare a first class proposal. With some encouragement from other organizations and with some internal soul searching we made the commitment. In June of 1992 we were informed that the Prince Albert Model Forest proposal had been successful. When we signed our formal agreement in early 1993 our partners were and remain:

- Saskatchewan Environment and Resource Management
- The Federation of Saskatchewan Indian Nations
- The Prince Albert Grand Council
- The Montreal Lake Cree Nation
- The Canadian Institute of Forestry
- Prince Albert National Park
- Weyerhaeuser Canada.

Before I go on any further with a more detailed discussion of the reasons for our decision to participate in this project and the benefits to Weyerhaeuser that we anticipate, I would like to provide you with one perspective of past forest resource management in Saskatchewan. Among other things, the Prince Albert Model Forest is a collaborative exercise, and a brief discussion of the past may help to get the Model Forest into perspective.

One of the first forest management agreements in Saskatchewan was signed in 1965 between the Province of Saskatchewan and Parsons and Whitemore of New York, with the subsequent construction of the Prince Albert pulp mill. That pulp mill was later purchased by Weyerhaeuser along with the Big River sawmill. This agreement and the mode of operations back in the 1960s, '70s and even into the early '80s was one of single resource management. The company had a "forest;" in those days meaning a "timber" management agreement. That was what we were primarily interested in. Management of other resources such as wildlife and integration of other users needs was the responsibility of the province and had little to do with us. On the surface this approach seemed to be clean, cost effective, and the right thing to do at the time. It is not to say that we totally shut out other users and gave no consideration at all, but delivery of wood at the cheapest price was our prime objective. It is probably safe to say that decisions were not always made in the best interest of all forest resources.

With respect to First Nations people and specifically Montreal Lake Cree Nation, a relationship developed with the new pulp mill in the late '60s and early '70s. A working area was set up specifically for Montreal Lake and they were considered to be a primary contract source for wood and for reforestation activities; and for a while they were. However, for a variety of reasons, the working relationship fell apart during the mid to late 1970s and for almost 15 years the relationship of the pulp company and the people of Montreal Lake was virtually non-existent with the exception of some minor reforestation contracting; but even that had some rough going. Why the relationship broke up is still subject to various interpretations depending on who you talk to. I think people tried to make it work. However, failure to identify, or recognize cultural differences and failure, or inability of the organizations of the day to adjust for those differences was probably one of the root causes.

In the last half of the 1980s and into the 1990s, as I'm sure you are aware, other changes were occurring within the resource management field as a result of changes in attitudes by citizens of this country and around the world. The environment and all of its ecological systems were subject to an awakening of concern by people; not only those who were familiar with resources and used them, but by a larger population of people who wished to be comforted in knowing that all species of life would continue to exist and that the overall health of the planet would continue.

A change in attitude started within our industry in the early 1980s. By way of example, our company contracted a project in 1985 to quantify the differences in animal and plant populations following harvesting operations versus fire. This project spurred on another project in 1989 called the Saskatchewan Forest Habitat Project with the objectives of finding ways and developing the tools necessary to integrate timber and wildlife management. This project was a true partnership of resource management agencies, governments, and industry. Several of the partners in this project also became partners in the Model Forest. The significance of this project I feel was that it represented some of the first concrete efforts in Saskatchewan to achieve integrated resource management in the forest. This was truly a commitment by all partners to that concept. This cooperative arrangement helped set the stage for the trust that was required for the Prince Albert Model Forest partnership.

However, in the early stages of putting together a proposal for a Model Forest, there was still a fair amount of relationship building and trust required between industry, government, and First Nations people to bring together a proposal and to be successful as a "Model Forest." I believe that the process of putting together a comprehensive proposal in a relatively short time frame helped to solidify our Model Forest partnership.

Having provided you with a very brief sketch of the past 30 years of forest resource management from a company perspective I hope that you can better appreciate why Weyerhaeuser decided that participation in the Prince Albert Model Forest would be a positive step towards the future.

Our reasons for joining the Prince Albert Model Forest partnership were essentially two-fold.

First, to continue to build on partnerships and relationship already initiated, recognizing that in today's society the involvement of affected stakeholders is essential in maintaining access to public resources for our mills. More specifically, we felt that it was absolutely essential for us to rebuild the relationship that had existed with Montreal Lake and to build an even stronger relationship than before based on not only an economic relationship but one in which the cultural and societal needs of the people of Montreal Lake were recognized and incorporated into considerations of where and when our operations occur on the license agreement area.

With respect to the second reason for our involvement in the project, I have talked about society's sensitivity to the environment and what is going on around us. People want to know that the impacts of our activities on the environment are sustainable and that our forest practices will not degrade the environment. To address these concerns that society, our governments and indeed ourselves have, we need good sound information. Therefore, we saw this cooperative project as an opportunity to improve on the overall knowledge base of the forest of Saskatchewan, with the ultimate goal of making better informed integrated management decisions.

CURRENT PERSPECTIVE

So where are we at today?

First, I would like to point out that the Prince Albert Model Forest was founded on the principles of an equal partnership. Our partnership includes industry, Federal Government, Provincial Government, three levels of First Nation government and a non-profit professional organization. However, all partners have equal status and are given equal voice. At the board of directors level all decisions are based on consensus after serious review by the Technical Committee and the Board of Directors. We discuss issues and differences and come to a unanimous agreement before formalizing our decision with a vote. The key is open and honest communications and common purpose and objectives.

What else has happened?

Weyerhaeuser's wood harvesting contracts with First Nations people have grown to in excess of 144,000 m³ of wood per year from about 27,000 m³ back in 1992. Montreal Lake Cree Nation is a full time wood supply contractor and a partner in another First Nation contracting company called "Woods Cree." Combined, these two contractors alone produced 44,000 m³ in our 1994-95 operating season. Reforestation contracting with Montreal Lake has grown and stabilized at a point where we are comfortable with guaranteed numbers of trees to plant and hectares to clean and thin on an annual basis. They do tree improvement contracting with us as well. In my opinion we have the foundation of a long-term relationship based on trust. Montreal Lake is a long-term contracting partner of Weyerhaeuser.

Montreal Lake and other users of the forest have also been involved in collaborative planning efforts for future harvesting areas. On a significant land area within the Model Forest area, our planners sat down with the people of Montreal Lake, with government biologists and foresters, and National Parks people and designed harvest blocks to best meet everyone's needs.

We are identifying areas of cultural significance on our license area, and will take appropriate action as our operations approach these areas.

We have implemented a summer student program over the last three years targeted at young people of aboriginal ancestry. Our hope is that we can encourage these young people to make a career choice of resource management and hopefully find employment with us in the future. We continue to have need for trained resource management professionals and technicians.

I believe that our relationships with other partners have continued to grow. However, we had the most to gain from our relationship building with First Nations, and to that I think we have grown the most.

With respect to our objective of making better management decisions, in the areas of research and technology transfer, the effort over the last two-and-one-half years are starting to bear fruit. Building on work started by the Saskatchewan Forest Habitat Project, Weyerhaeuser is now incorporating single leave trees and clumps of trees into our clearcut harvesting patterns to provide future habitat for birds and small mammals.

Additional information relative to biodiversity needs is forthcoming and we will be incorporating that into our forest management plans in the future. Workshops on computer modelling have provided insight into options that are available to us as we begin to more closely assess the available wood supply to our mills in light of other resource needs.

The G.I.S. literacy program initiated by the Model Forest in cooperation with our local technical training institution will result in skilled G.I.S. technicians in the near future. As I speak young people of aboriginal ancestry are working side by side with our G.I.S. and G.P.S. technicians, learning the technology and contributing to our business as they complete their practicum.

CLOSING REMARKS

Today we have less than two years remaining on what was originally a five-year project. We are confident that this partnership can continue beyond the original term. We feel that the trust that has developed through this working partnership will ensure that. It can only be to our mutual benefit. Weyerhaeuser has a vision of future forest resource management across our entire license area and beyond, based on the same principles and cooperation that has developed through the Model Forest. The Model Forest organization could very easily become the central repository of resource information and practical forest resource research for all of northern Saskatchewan, providing a service and filling a gap that will be created by the recently announced reduction in our federal forest service. Industry has a need for good decision making information as do all agencies involved in resource management.

Weyerhaeuser looks forward to the upcoming years as we forge new relationships and build on old ones with the people who live, work and manage the vast natural resource of northern Saskatchewan. At some point in time in the future I believe people will look back and see the Prince Albert Model Forest as a key turning point for resource management in this province.

Thank you.

Precious Values:
Integrating Diverse Forest Values
into Forest Management Policy
and Action (Ontario)

Mark A. Stevenson
David R. Hardy
Laurie Gravelines

ABSTRACT. Public participation has wide support among various stakeholder groups as a means of identifying diverse forest values. However, attempts to incorporate these values–especially non-commercial values–into decision-making have not been successful. The Non-Commercial Forest Values Colloquium was sponsored by the Ontario Ministry of Natural Resources to redress this omission. Looking at a range of non-commercial values, the colloquium developed a set of Core Principles, viz.: respect the totality of interests embodied in a forest; adopt a holistic ecosystem perspective; guide forest policy through a set of Core Principles; give primacy to the preservation of ecosystem integrity; avoid irrecoverable harm in policy decisions; and develop accounting/evaluation systems that balance values in forest policy decision-making. *[Article copies available for a fee from The Haworth Document Delivery Service: 1-800-342-9678. E-mail address: getinfo@haworth.com]*

Mark A. Stevenson and David R. Hardy are Principals of Hardy Stevenson and Associates Limited, Toronto, Ontario, Canada.

Laurie Gravelines is Forest Values Project Manager with the Ontario Ministry of Natural Resources, Saulte Ste. Marie, Ontario, Canada.

[Haworth co-indexing entry note]: "Precious Values: Integrating Diverse Forest Values into Forest Management Policy and Action (Ontario)." Stevenson, Mark A., David R. Hardy, and Laurie Gravelines. Co-published simultaneously in *Journal of Sustainable Forestry* (Food Products Press, an imprint of The Haworth Press, Inc.) Vol. 4, No. 3/4, 1997, pp. 171-183; and: *Sustainable Forests: Global Challenges and Local Solutions* (ed: O. Thomas Bouman, and David G. Brand) Food Products Press, an imprint of The Haworth Press, Inc., 1997, pp. 171-183. Single or multiple copies of this article are available for a fee from The Haworth Document Delivery Service [1-800-342-9678, 9:00 a.m. - 5:00 p.m. (EST). E-mail address: getinfo@haworth.com].

171

INTRODUCTION

The future of Canada's forests has become a major environmental issue. Witness the disputes throughout Canada: in British Columbia regarding logging in Clayoquot Sound; in Alberta over the allocation of northern forests to pulpwood production; and in Ontario with the protection of old growth forests. What we are seeing is the conflict between people holding diverse values and strong opinions about the meaning and the future of the forest. There is agreement in principle among industry, government, interest groups and the public on the importance of identifying the values through public participation and the need to address all forest values in policy decisions and land allocations. However, in practice, integrating diverse forest values into these decisions continues to be elusive.

In an attempt to move beyond the conflict over values, the Ontario Ministry of Natural Resources' (OMNR) Forest Values Project initiated a comprehensive approach to forest values by remaining open to consideration of any values that the public, industry and interest groups felt it should address. While, in principle, this openness may help to reduce conflict, it also highlights two persistent and major weaknesses with its standard approaches to forest valuation. First, how do we value non-commercial values of the forest (e.g., intrinsic values) or uses for future generations (e.g., option values). Second, how do we ensure that these values represent a comprehensive range of values which can contribute to an informed decision-making process.

The first task OMNR undertook to understand these forest values was to consult with the public about what values are relevant and what mechanisms would be suitable for widening their consideration. The advice from the 1993 public consultation efforts with over 100 groups, including foresters and forest industry representatives, was clear but less specific than expected. The groups confirmed that OMNR was on the right track, stating that OMNR should take a closer look at how it ought to consider non-commercial forest values. And, they suggested that OMNR develop a position paper addressing several key questions, including:

- Are we properly valuing our forest resources?
- How should we best approach these values in our forest decision-making?
- What tradeoffs must be made to provide a valuation of non-commercial values for consideration in forest policy and management decisions?

In response, the Ministry sponsored a "Non-Commercial Forest Values Colloquium" in September 1994, at the University of Toronto and produced a report: "Precious Values" (Hardy Stevenson and Associates Ltd. 1994). This paper focuses on the results of the Colloquium.[1]

WHAT ARE NON-COMMERCIAL FOREST VALUES?

We are very familiar with commercial values. They are values assigned to our forests based on forest industry activities (e.g., timber, pulp and paper) and non-industry commercial activities such as hunting and tourism. We can quantify these values and assign a dollar value. However, at best, the discussion of commercial values represents just half of the public policy debate.

Background research for the Colloquium confirmed that there is a range of non-commercial values; values attributable to the forest that are not easily measured or quantifiable in dollar terms but ought to be explicitly considered in forest policy decisions and actions. However, addressing all current forest values is important to ensure that we are making wise and sustainable policy choices. These values represent the other half of the public policy debate. They include but are not limited to:

- *Intrinsic Values*–the value something has in itself, independent of its value to any other being. For example, if something can be damaged or harmed, it has some value, and this value is independent of, for example, its usefulness to other beings;
- *Spiritual Values*–the value attached to something *by people because of its* importance to their spiritual and cultural sense of identity (e.g., special relationship of Aboriginal people to the land). It would also include our aesthetic response to the forest;

- *Ecological Values*–the value attached to the importance of maintaining forest bio-diversity to ensure overall ecosystem health and human survival;
- *Community Values*–this value is central to understanding the sense of community identity, community cohesion and the quality of life in many forest communities;
- *Existence Value*–the value attached to the satisfaction people obtain from an amenity for various reasons other than their expected personal use; the satisfaction or peace of mind which comes just from knowing that a natural feature, such as a forest, exists independently of our experience or use of it.

There was a great deal of discussion at the Colloquium on how to classify and define non-commercial values. The group felt that a classification was helpful in thinking through the issues.

THE COLLOQUIUM RESPONSE

How should we address the wide range of values that are at the root of the conflict in forest policy decision making? The results of the colloquium provided pragmatic advice to forest policy makers and planners.

Recognize and Respect the Principle That the Forest Represents a Diversity and a Totality of Interests and Embodies a Spectrum of Values

The first problem tackled by the Colloquium was our tendency to look at the forest from our own personal or professional set of values. We tend to look at the forest this way because our personal or professional values are, to each one of us, our fundamental beliefs that guide all our actions. Where the conflict in forest policy occurs is when one set of values has, or appears to have a disproportionate role in the outcomes of forest policy. Or when one value is omitted from consideration. To address this, the Colloquium concluded that the forest should be viewed as representing a spectrum of values including tourism, mining, water, existence, intrinsic and other values. And, forest policy must clearly demonstrate that it has understood and addressed this spectrum of values.

However, Colloquium participants felt that while the principle of treating all values with equal respect–as integral parts of a "spectrum of values"–is important, it does not sufficiently address the problem of how to account for the essential "ecosystem values" of the forest.

Emphasize That Policy Decisions on Forest Management Should Be Made Within a Holistic Ecosystem Perspective

Colloquium participants told us that looking at forests from an ecosystem perspective involves the consideration of three major and inter-related components: natural systems, socio-political systems and economic systems. Like natural systems, both social and economic systems have their own sub-systems.

In considering social systems, we need to shift our perspective to consider that the decision to cut a tree, plant a tree or change a species has implications for people working in the forest, for storekeepers who depend on the forest worker's spending, for children, teachers, and other people in the local and wider community. The forest social system encompasses other systems related to housing, social and community services, transportation, retail services, etc. And, a forest policy decision affects this component of the broader ecosystem.

In addition, they also stressed the importance of recognizing that commercial values are integrated with non-commercial values. Colloquium participants felt that the OMNR would continue to experience difficulties if it attempted to incorporate non-commercial forest values into their policy framework by carrying on with an indirect process which isolates non-commercial values as if they were somehow separate from commercial decisions. Colloquium participants looked at this problem and stated that dealing meaningfully with non-commercial values is, therefore, a "process" issue of the first order.

As one way of addressing this, they stressed that the forest policy decision-making process must not be seen as a "linear, utilitarian process" in which the Ministry just adds on some non-commercial values to the existing process. It may, for instance, actually be harmful to the forest policy process to persist with a system in which particular stakeholder groups approach the Ministry and "invoke the intrinsic value" of a specific geographic area as an "add-on" to the current process. Instead, the articulation of non-commercial values must be a normally integrated part of the forest policy evaluation process–guided by Core Principles.

Guide Forest Policy Decisions Through Core Principles Aimed at Preserving Ecosystem Integrity

Traditionally, we tend to look at forest decision making as a rational sort of process. Where decisions must be weighed and evaluated, we tend to think about what outcomes will maximize the interests of the broader public interest and common good. This is often called, the *utilitarian approach*. And, it is this approach that was seen to be a problem for addressing non-commercial forest values.

Colloquium participants stated that to address non-commercial forest values, the use of the utilitarian approach to determine industrial values requires supplementation by distributionally sensitive principles that would, for example: incorporate the values of those whose ways of life are most strongly affected by economic and forest use decisions; give weight to the consideration of all living things; avoid irrecoverable harm; and recognize the effects of contemporary policy decisions upon future generations. However, Colloquium participants went further and suggested that what would be better than an ad hoc consideration of the interests of people supporting one or another non-commercial values would be a set of Core Principles helated to a range of forest policies or area of land–agreed to in advance.

Specifically, Colloquium participants felt that the development and respect for Core Principles would assist in fully integrating the broad spectrum of values. Some principles could be considered as first principles, such as long-term ecosystem integrity and the welfare of future generations. Establishing such Core Principles would assist in focusing the discussion on the values at work in policy decisions. As a consequence, the public discussion should begin by focusing on what shared principles are essential.

The Primary Principle Should Be to Sustain Ecological Integrity

These Core Principles can also function to protect intrinsic values. For example, the principle of preserving ecosystem integrity can help to address intrinsic values. Rather than identify specific areas of intrinsic value, we should recognize that intrinsic value applies to everything and preserving ecosystem integrity may best protect this value.

Core Principles must also be monitored to check their relevancy. Emerging values and emerging world views related to the forest must be understood, compared to Core Principles and probed as a potentially important input to forest policy.

Adhere to the Principle of Avoiding Irrecoverable Harm in Policy Decisions

Colloquium participants suggested that a principle of avoiding irrecoverable harm should be applied to our uses of the forest to assist us in making wise choices in accordance with the guiding principle of not taking actions that would jeopardize ecosystem integrity. Particularly relevant for biodiversity, it means avoiding actions that permanently destroy or eliminate anything.

This principle assumes that comprehensive harm and benefit analyses would be applied to all of our forest activities. This analysis would combine, for example, quantitative analyses and studies on values preferences. It would then be necessary to try to link these quantifiable sources of data with qualitative data and judgements about forest values.

At the same time, adopting the principle of "Avoiding Irrecoverable Harm" will respect the "Unknown and the Future Values" of the forest by both ensuring that the changes resulting from our forest activities are reversible, and that unknown or potential future values are protected.

In addition, it provides guidance to forestry policy makers when we say: "avoid irreversible harm," "protect biodiversity" and "respect future values." Thus, when specific forest decisions are made, we are effectively saying, "do not override these principles unless you have a very good reason for doing so."

Develop Accounting and Evaluation Systems to Make Tradeoffs
and Balance Values in Support of Forest Policy Decision-Making

Processes and Core Principles are fine as far as they go, however, the problem for most foresters remains: how do we implement these principles? What are some of the more immediate and tangible steps we can take to improve forest decision making as it related to non-commercial forest values—particularly those pressing eco-system values? Ultimately, a decision-making process which hopes to take account of a wide range of non-commercial forest values must acknowledge that some of these values are difficult, if not impossible, to quantify. For example, how is it possible to quantify the damage to or loss of a forest in terms of its spiritual significance to people? A process for incorporating qualitative values is still required.

Colloquium participants addressed this and suggested that we must develop an "integrated framework" that accommodates a way to grant value, other than "use values," to things in their own right irrespective of any use we might make of them. For example, personal interpretations—such as a child's rendition of what he or she values about the natural environment—are valid in this discussion, but they are currently not accommodated well in our public policy processes.

THIRD AGE ACCOUNTING
AND OTHER EVALUATION METHODS

The Colloquium confirmed Third Age Accounting as one of the key tools that can be useful in supporting the analysis of non-commercial values and the decision making process. Third Age Accounting, like many other tools, was developed to support decision-makers. Essentially, it is a broad forestry planning balance sheet where all commercial and non-commercial values are considered. It is called "Third Age Accounting" for the following reasons.

First Age Accounting is the simple cash accounting ledger and essentially pre-dates capitalism. Decision-makers simply needed to know what was left at the end of the day to survive. Second Age

Accounting or accrual accounting was developed with the emergence of capitalism. Decision-makers needed an accounting framework that could deal with the investment of capital at one time and then consumed over a long period of time.

Third Age Accounting is emerging because decision-makers need a broader range of information. They require more information on the impact of their options on people, the environment and on use and non-use values. Thus, the Colloquium pointed to the need to explore the development of a holistic accounting framework.

The usefulness of *contingent valuation* within this accounting system was also discussed. Contingent valuation is the use of surveys or other methods of measuring or mapping to quantify how much value people attribute to a thing. The use of criteria, indicators and data arising from contingent valuation techniques gives credibility to forest values decisions because the decision is based on quantitative information. However, it overlooks the question of how much information alone can decide for us compared to the essential role of judgement in assessing the relative strengths of various values. Still, participants felt it would still be useful to derive measurements of the importance of various values to the public, even if we concede that we cannot quantifiably measure the values themselves.

Consider Institutional and Structural Change Designed to Integrate Provincial and Local Communities as Decision Making Partners

The Colloquium participants strongly agreed with the concept of local community empowerment. However, they also felt strongly that community empowerment related to forestry decisions must occur within a provincial framework.

They felt that the provincial interest in key areas such as economic strength, the social well-being of Ontario residents and in environmental protection should remain significant policy determinants; with ecosystem health being primary. Essentially, the Ontario Ministry of Natural Resources' mandate to reflect province-wide—and indeed global—values, should not change.

Subject to the Province retaining overall management of forest lands, local communities should be empowered to make important decisions which affect them. They should be entrusted to identify and specify where non-commercial values should be at work. This means that the Ministry's role as "manager" does not equate with having complete control; rather, it means establishing the broader framework within which community-based decision-making can take place. However, all Colloquium participants did not agree.

Local Decision-Making Must Be Free of Overriding Economic Pressures

Two pressures operate at a global level and place great stress on resource communities:

1. the global economic system continues to mine resources and does not promote their sustainability. This sustainability issue is underpinned by assumptions about current approaches to economic development, work and the distribution of societal wealth; assumptions that affect, but also transcend, forestry policy decision making; and,
2. the population explosion with its ever-increasing demand for resources will increase the demands on the forest resource. For example, European tourists now pay considerable sums to experience Ontario's forest resources at the same time that the demand for wood products continues to grow.

To address these pressures, participants felt that there is a need for more adaptive and comprehensive strategies for making decisions at the community level.

Ideally, there should not have to be a trade-off between destroying the local resource base on the one hand and the survival of the community on the other. To address this, the Province needs to seriously examine opportunities for diversifying forest-based economies in order to ensure the survival of community values. Participants felt that one way to "level the playing field" is to provide some form of compensation for local communities. Where local knowledge concludes that non-commercial values of an area must be protected, communities must be supported to make the decision without fear of economic consequences.

PLAN FOR THE LONG-TERM

Colloquium participants felt that the aboriginal tradition in Ontario of thinking about a seven generation time span is a helpful model for shaping this broader perspective. Over the short-term, we must be able to change practices in light of new knowledge and the assessment of impacts of present practices. However, we must accept that since we don't know all of the implications of our actions and our level of knowledge is incomplete or unknown, our time horizon for preserving ecological integrity must be long-term, that is, seven generations into the future.

CONCLUSIONS

Colloquium participants concluded that a number of important approaches could be used to improve forest policy decision-making pertaining to incorporating non-commercial values. The first is the need to recognize the complexity involved in the discussion of non-commercial values. Many values are at play in forest policy decisions. To begin to improve our approaches, we must be clear about how we define non-commercial values. Participants felt that the differentiation of commercial values versus non-commercial values may be more artificial than real. Such a differentiation creates the opportunity for the hardening of positions among actors within forest policy discussions. Having better information about all of the values at play will begin to clarify these decisions.

The second approach involves two aspects of change to the forest decision-making process. Initially, there is a need to acknowledge the central role of a set of Core Principles developed by all participants to guide decisions. In addition, there is a need to develop new accounting and evaluation frameworks to recognize the holistic nature of forest values. This includes the need to recognize existing values and probe emerging values. Core Principles and the accounting framework will provide the benchmarks against which government can be held accountable for its ecosystem management and the management of resources.

The third, and perhaps the strongest approach, involves the need to focus on changing the processes within which non-commercial values are addressed in forest decision-making. The change involves moving decision-making closer to local communities. In essence, there are fundamental shifts required in our perspectives on forest values which have to be reflected in structural changes to Ministry policy.

While specific trade-offs were not addressed during the course of the Colloquium, adoption of the overall guidelines for forest decision-making imply a host of practical ways to approach those trade-offs, e.g., as a society, re-examining our uses of the forest to see if they are as important as we think they are–given the wide range of values we impact through our activities.

However, the participants emphasized that the dialogue needs to continue. More research and discussion is required on:

a. a framework and consultation process for dealing more adequately with the interrelationships of commercial and non-commercial forest values;

b. the range and importance of non-commercial forest values currently held by the people of Ontario;

c. criteria and indicators for sustainable forestry practices;

d. forest accounting frameworks which incorporate "qualitative data" on values and methodologies for factoring in intergenerational liabilities.

The Ontario Ministry of Natural Resources has taken some critical steps in recognizing that non-commercial values are wide-ranging and important, and that methodologies for incorporating these values into forest decision-making processes are required. Consultation processes are central. This process is well underway and there are grounds for optimism.

NOTE

1. The participants in this Colloquium included: an Aboriginal spiritual leader and an Aboriginal forest policy advisor, an Environmental Studies professor, philosophers, economists, an accountant, a theologian, foresters, and a representative of Northern Ontario local municipalities. They are all experts in different fields and were asked to address the problem of non-commercial values from the perspective of their own discipline.

REFERENCE

Hardy Stevenson and Associates Ltd., Precious Values: A Report by the Colloquium on Non-Commercial Forest Values. Ontario Ministry of Natural Resources, December, 1994.

Socio-Economic Sustainability in Forest Dependent Communities

Phil Shantz

ABSTRACT. This paper has as its emphasis, the identification, description and exploration of the relationship between timber supply, forest industries and regional and community socio-economic sustainability in Ontario. Six issues have been identified for discussion: (1) the socio-economic dependency of communities and regions on the forest industry; (2) the socio-economic effects of varying levels of timber supply on the forest industries and communities and regions in seven case study regions; (3) typical socio-economic effects when various levels of timber supply are either added or reduced from a region, while identifying potential mitigation and enhancement measures and net effects; (4) factors influencing the socio-economic sustainability of forest-based communities; (5) a framework, criteria and indicators for assessing the sustainability of forest-based communities and industries; and (6) how resource management and socio-economic analysis and planning can promote community sustainability. *[Article copies available for a fee from The Haworth Document Delivery Service: 1-800-342-9678. E-mail address: getinfo@haworth.com]*

INTRODUCTION

The forest industry is a major contributor to the economy of Ontario. It directly employs approximately 70,000 people in log-

Phil Shantz is Consulting Planner/Analyst for the Engel Consulting Group.

This project was carried out for the Ontario Ministry of Natural Resources when he was with Hardy Stevenson and Associates Limited, Toronto, Ontario, Canada.

[Haworth co-indexing entry note]: "Socio-Economic Sustainability in Forest Dependent Communities." Shantz, Phil. Co-published simultaneously in *Journal of Sustainable Forestry* (Food Products Press, an imprint of The Haworth Press, Inc.) Vol. 4, No. 3/4, 1997, pp. 185-196; and: *Sustainable Forests: Global Challenges and Local Solutions* (ed: O. Thomas Bouman, and David G. Brand) Food Products Press, an imprint of The Haworth Press, Inc., 1997, pp. 185-196. Single or multiple copies of this article are available for a fee from The Haworth Document Delivery Service [1-800-342-9678, 9:00 a.m. - 5:00 p.m. (EST). E-mail address: getinfo@haworth.com].

185

ging, forestry services and primary and secondary wood using industries and one job in thirty-one in the province is dependent on the industry (Forestry Canada 1994). The economic benefits from the primary wood using industries are derived from harvesting operations and approximately one hundred large processing facilities throughout the province which produce lumber, pulp, paper, composite boards, plywood and other timber based products.

While the industry is distributed throughout the province, most of it is concentrated in the north where almost fifty communities are significantly dependent on the industry for their economic livelihood. In Northern Ontario, almost two-thirds of the population lives in a community where the forest industry is a factor of some significance in the local economy and is part of the social fabric of the community. Because of their high level of economic dependence on the forest products industry, many communities within the province are significantly influenced by forest resource policies, particularly when these policies lead to changes in timber supply.

ONTARIO'S FOREST MANAGEMENT PLANNING SYSTEM AND SOCIO-ECONOMIC ANALYSIS

Ontario is an important timber producing region. The forest products industry in Ontario produces approximately 3.5 million tonnes of paper and paper board, 1.6 million tonnes of market pulp, 6 million cubic meters of lumber (95% softwood lumber) and 1.0 million cubic meters of board and veneer. Annual industrial consumption varies between 20 and 25 million cubic meters of wood fibre annually (Callaghan 1994). This wood fibre originates primarily from the northwestern, northeastern and central parts of the province.

This timber producing area of the province is segmented into approximately one hundred different Forest Management Units (FMUs) and managed by the Ontario Ministry of Natural Resources (OMNR) for forest management.[1] Each of these forest management units has a plan prepared for it. Extensive socio-economic and cost/benefit analysis have been absent from these plans. Various provincial stakeholders, in particular representatives from forest industry dependent communities, have identified that the failure to analyze how communities are affected by resource management decisions is a serious weakness in forest management planning (Hardy Stevenson and Associates 1993). In response, the OMNR (1994) has stated within its *Policy Framework For Sustainable Forests*, that the long-term viability and sustainability of forest-based communities and businesses are vital factors, second only to the sustainability of forest ecosystems.

One of the main initiatives to address socio-economic issues in forest dependent communities, was the study, *Timber Supply and Community Socio-Economic Sustainability*, which was undertaken by Hardy Stevenson and Associates Ltd. (1995), for the Forest Resource Assessment Policy Project (FRAP). This study was undertaken in order to: identify and describe the factors that affect and contribute to the socio-economic sustainability of forest-based communities; identify steps that can be undertaken to promote community socio-economic sustainability; and, identify how the OMNR can adjust its land use and resource management planning system to promote their sustainability. This study employed a literature review and field research on seven case studies. The case studies were: the Algonquin Forestry Authority; the Eastern Ontario Model Forest; the Elk Lake Community Forest; a study of one integrated forest products company–Malette Inc. with its operations in Timmins and Smooth Rock Falls; Fort Frances District; the logging communities of Red Lake and Ear Falls; and, Sioux Lookout District. These case studies represented different forest types, industrial structures and regions and communities throughout Ontario. One of the main products of this research and the focus of this paper, was the identification of factors influencing forest dependent communities.

FACTORS INFLUENCING ONTARIO'S FOREST DEPENDENT COMMUNITIES

Six factors influencing the socio-economic sustainability of forest industry dependent communities were identified. These included: nature of the local and regional economy; demand for forest products and provincial and regional timber supply trends; local forest products industry factors; land tenure and licensing issues; community social factors; and, labour mobility.

Nature of the Local and Regional Economy of Forest Dependent Communities

The economy of Northern Ontario is resource based and will continue to be resource based in the foreseeable future (Advisory Committee on Resource Dependent Communities in Northern Ontario 1986). The development of communities and regions has been a spatial reflection of the resource based economic structure. Communities and industries have developed in close proximity to the resource because of the transportation costs of the pre-processed raw material compared to the processed commodity. The resource industries in the north: forest products; mining; energy; and, tourism are the basic economic sectors upon which Northern Ontario communities are dependent.

A study comparing Northern Ontario forest dependent to non-forest dependent communities drew the following conclusions: "(1) Forest dependent communities in the north tend to be wealthier than non-forest dependent communities; (2) Forest dependent communities generally tended to have reduced incidence of low income earners; (3) The greater the dependence on the forest products industry in the community, the lower the percentage income earned from government transfer payments by individuals; and, (4) Forest dependent communities in the north tended to be somewhat healthier than non-forest dependent communities in terms of unemployment rates and labour force participation rates" (C.N. Watson and Associates 1989). However, all forest industry dependent communities are not the same and an understanding of their differences will lead to improved decision-making for forest resource planning and economic development. Forest dependent communities can be classified into four categories.

The first category is harvesting communities, that are involved in timber harvesting operations and have very limited or non-existent primary wood using industries. These communities are poorer and have less stable populations than communities with large mills.

The second category is comprised of processing communities where the local industry processes timber into products such as lumber, board and/or pulp and paper. A smaller portion of the labour force may also be involved in logging. For the most part these communities are generally wealthier than other Northern Ontario communities, in particular if a large unionized pulp and paper mill is present. The high wages paid by the large companies to the unionized labour force has often acted as a disincentive to economic development in these communities. Many of these communities are highly or solely dependent on the local mill and would be dramatically affected by a mill shutdown or closure.

The third category of communities are the regional service centres. These communities may have industries that harvest and/or process timber, but they have evolved into regional economic centres with well developed retail, construction, business service and government sectors. These communities have grown in response to the regional expansion of the resource industries. From the perspective of regional economic theory, they are not growth poles, quite the contrary they generate few economic effects to their peripheral regions. Rather, the smaller communities that harvest and process natural resources provide significant economic benefits to these larger communities. A significant proportion of their economic base is dependent on the viability of smaller communities and the economic development strategies of these communities should reflect this regional dynamic.

The fourth category is comprised of bedroom communities where residents are employed outside municipal boundaries. These are small communities with populations of generally under five hundred. These communities may be attractive communities for living, have a significant percentage of the population that is either retired or on social assistance and/or may have had a large employer that has either closed, downsized or relocated. These are poorer communities with a tax base highly dependent on residential assessment.

Demand for Forest Products and Provincial and Regional Timber Supply

Historically, forest industry dependent communities in Ontario have been affected more by demand and competitiveness factors rather than timber supply issues. However, timber supply is becoming the more important issue that will determine the stability of Ontario's forest products industries and forest dependent communities, as there is increased pressure in the province to withdraw land from the industrial timber landbase in order to preserve more land for ecological, wilderness and recreational purposes.

Through the *Forest Resource Assessment Policy Project*, the Province of Ontario has recently completed one of the most comprehensive timber supply and forest products demand projects.[2] This project has been important in the identification of timber supply opportunities and problems. One of the most practical outcomes of the project was the recognition that the province's intolerant hardwood forest has been a significantly underutilized resource. Less than three million cubic meters of hardwood fibre were used by the industry, while supply was greater than ten million (Callaghan 1994). This supply together with the opportunities that exist in the marketplace[3] have led to the establishment of three new oriented standboard mills in the province, the expansion of another and the development of a specialty hardwood mill. This represents one of the largest industrial developments in Northern Ontario in many years.

The FRAP project has also identified that while the softwood timber supply currently exceeds demand, this will change in twenty to thirty years (OMNR, 1994). Some areas of the province with a high dependence on softwood will experience the impact more severely, in particular if new protected areas are established and woodflow regimes cannot be shifted. The province can mitigate many of the potential economic impacts of this trend, but it must implement the mitigation strategy as soon as possible.

Local Forest Products Industry Factors

If forest products industries are to remain a key cornerstone of the economies of Northern Ontario communities, the industries must remain competitive. Numerous factors affect the competitiveness of forest products companies and employment opportunities in forest based communities. These include: timber supply; fibre costs; labour productivity; good labour-management relations; location of manufacturing; environmental and safety regulations; implementation of modern technology and practices; energy costs; good management; achievement of economies of scale; access to capital;

government's regulatory and policy framework; ability to develop new products; and, product demand.

Forest products are commodities whose prices are determined on continental and world markets. As national economies give way to integrated continental and global economies, commodity producers must strive to be cost competitive. Similar to other manufacturing industries; increases in production are being made by substituting technology for labour. While direct employment in the industry is decreasing per volume of fibre consumed, indirect employment has increased as mills require new technology. Although the majority of this technological development will not occur in timber producing regions, but rather accrue to industrial areas.

The gradual erosion of employment per volume of fibre consumed has resulted in a great deal of criticism on the forest products industry. One of Ontario's main environmental groups, the *Wildlands League*, has initiated the *Forest Diversity–Community Survival Project* (1995). This project has developed several newsletters raising concerns about the smaller proportion of jobs per volume of wood consumed in the forest products industry.

An issue salient to this debate is whether the province is receiving the maximum economic value from the wood resource. There exist numerous opportunities, particularly in the southern areas of the province, where there is a higher quality wood resource, to undertake more value added production and to ensure that the higher quality wood resources go to value added operations. Many smaller companies and mills, local entrepreneurs and small industry advocates have been spearheading these efforts. The Province should be actively supporting such efforts as these value-added operations would provide significant and realistic economic diversification for many forest dependent communities.

While there are numerous structural trends apparent within Ontario's forest products industry, two in particular deserve attention. First, there has been significant growth over the last twenty years in regionally based forest products companies. In both the Northeast and Northwest Regions, Ontario based companies have been successful by purchasing older mills, modernizing them and by running very cost effective operations. These modernized mills have often been located in economically undiversified communities and represent the only major industry in these communities. These regionally-based companies often do not have organized labour in the mills and very rarely within woodlands operations. The second trend within the sawmill sector, is the movement towards larger mills, maximizing both labour and wood efficiencies. Many of the softwood lumber mills are taking smaller logs, operating on three shifts per day and are producing greater than one hundred million board feet per annum.

Land Tenure and Licensing

The current land tenure and licensing situation significantly affects the wood supply of some companies and stability of many communities. Currently about seventy percent of the forest land licensed for harvest in the province is covered by Forest Management Agreements. Industry has a strong preference for this longer-term tenure arrangement.

Within the central region of the province there is a significantly different tenure structure than compared to the northeastern and northwestern regions. There are no large FMAs or Company Management Units and there is a trend in some areas of the central region to shorter term, small scale sales and licenses (Brown 1993). The lack of longer term tenure together with the gradual erosion of the industrial timber landbase is weakening confidence in the industry and causing capital formation problems. Longer term arrangements would add some stability to these companies and the communities in which they are located.

Some interest group leaders, particularly from Northern Ontario, have argued that communities should have a greater role to play in land tenure arrangements and forestry decision-making (Hardy Stevenson and Associates 1993). The Province has moved forward slowly to involve communities more in forest management by organizing local citizens committees. The Province's Community Forest Pilot Project has also led to increased community involvement in forestry (OMNR 1995). These have been important and necessary initiatives.

One of the outcomes of this research project, is that there is not likely to be one model of how communities should be involved in forest resource management issues. In the Mattawa region, located just to the north of Algonquin Park, community and industry representatives have organized to ensure the survival of the industry, develop value added production and in particular prevent any further timber landbase withdrawals. In northwestern Ontario, the logging communities of Red Lake and Ear Falls have argued for more local control of tenure and licensing so that their communities would receive greater economic benefits from the forest products industry. Finally, the Elk Lake Community Forest, faced with a variety of resource management issues and land use conflicts within the Temagami region, is identifying ways of resolving resource management conflicts to reflect local community needs while respecting the provincial interest (OMNR 1995). Gradual, but increased community involvement in forest resource management decision making is necessary in order to balance community interests with other interests.

Social Factors–Local Government, Leadership, Local Institutions

While few studies have documented the importance of community social institutions, local government and leadership to the sustainability of forest dependent communities, these factors play a vital role. They are important to the quality of life in communities, provision of social services, in developing the local economy and to support increased community involvement in resource management decision-making.

The importance of quality local government and local leadership on economic and resource management issues is essential if communities are to play a greater role in resource management decision-making. The Elk Lake Community Forest has played a significant role in local resource management issues because it has developed quality leadership, familiar with resource management and economic development issues. The OMNR could encourage greater local involvement in resource management in other areas by providing education and training opportunities for local leaders and participants in local citizens committees.

Labour Mobility

Over the long run, communities are likely to be more stable if the majority of employment opportunities are permanent rather than seasonal or temporary. Community sustainability appears to be better served by encouraging excess labour to move from communities with few economic prospects to regions which offer more.

The relocation of labour from less prosperous regions to more prosperous regions has often been considered one form of adjustment to the problems of underdeveloped regions (MacDonald Commission 1985) and to resource industries where the overall employment levels may need to be reduced. If the Province or any timber producing region is considering the establishment of large protected areas which will significantly affect timber supply, it is recommended that labour adjustment programs be considered as a possible mitigation measure.

SUMMARY

Forest dependent communities in Ontario and in other forest products producing regions have been and will increasingly be influenced by factors affecting the forest products industry. The narrow economic base of these communities means that they are very vulnerable to changes in technology, commodity price fluctuations, and increasingly, resource management plans and decisions. An understanding of how these communities are affected by these factors, together with a process to assess their impact and devise mitigation strategies; is essential if governments are to successfully implement sustainable forest management and ensure the socioeconomic sustainability of forest dependent communities.

NOTES

1. In Ontario there are three types of management units. *Crown Management Units* are where the OMNR prepares Forest Management Plans and may carry out timber management operations itself. The OMNR may also contract operations to individuals or companies or issue short-term licenses (up to five years) to companies which then carry out operations according to the approved plan prepared by OMNR. *Company Management Units* are where the unit is licensed to large forest companies which play a greater role in forest management. Planning, provision of access and harvest operations are all carried out by companies; however, the OMNR will normally carry out the activities of renewal and maintenance. *Forest*

Management Agreements are similar to Company Management Units in that these units are licensed to large companies. The major difference is that the companies have agreed, through negotiated agreements with the OMNR to carry out the planning, and all operational aspects of timber management except protection operations (Environmental Assessment Board 1994).

2. For a brief description of this project, see: Callaghan, Brian. 1994. *Assessing Ontario's Forests–A Status Report for the National Timber Supply Conference*. Kananaskis, Alberta.

3. To better understand the industrial demand for timber and to forecast roundwood demand at regional and sectoral levels an econometric model of Ontario's forest industry was developed and an industry analysis and forecast, *Ontario's Forest Products and Timber Resource Analysis* was prepared for the FRAP project by Resource Information Systems Incorporated (1992).

REFERENCES

Advisory Committee on Resource Dependent Communities in Northern Ontario, Ontario Government. 1986. Final Report and Recommendations of the Advisory Committee on Resource Dependent Communities in Northern Ontario. Ontario Government: Toronto.

Brown, W. J. 1993. Review of Wood Supply and Distribution The Southern Portion of the Central Region of The Ministry of Natural Resources. Huntsville.

Callaghan, Brian 1994. Assessing Ontario's Forests–A Status Report for the National Timber Supply Conference. Kananaskis, Alberta.

C. N. Watson and Associates Ltd. 1989. Socio-Economic Input To The Class Environmental Assessment For Timber Management On Crown Lands In Ontario. Toronto.

Davis, Craig H. 1990. Regional Economic Impact Analysis and Project Evaluation. University of B.C. Press: Vancouver.

Environmental Assessment Board. 1994. Reasons for Decision and Decision: Class Environmental Assessment by the Ministry of Natural Resources for Timber Management on Crown Lands In Ontario. Toronto: Environmental Assessment Board.

Forest Industries Action Group. 1993. Ontario's Forest Products Industry. Hard Choices–Bright Prospects.

Forestry Canada. 1994. The State of Canada's Forests 1993. Natural Resources Canada, Canadian Forest Service: Ottawa.

Hardy Stevenson and Associates Ltd. 1993. Timber Production Policy Options Development Workshop Report. Prepared for the Timber Production Policy Project.

Hardy Stevenson and Associates Ltd. 1995. Timber Supply and Community Socio-Economic Sustainability. Prepared for the Forest Resource Assessment Policy Project.

MacDonald Commission. 1985. The Royal Commission on the Economic Union and Development Prospects for Canada Report, Volume 3. Minister of Supply and Services Canada: Ottawa.

Markeusen, A.R. 1985. Profit Cycles Oligopoly and Regional Development. The MIT Press: London.

Ontario Ministry of Natural Resources. 1994. Policy Framework for Sustainable Forests.

Ontario Ministry of Natural Resources. 1994. The Forest Resource Assessment Policy. Sault Ste. Marie.

Ontario Ministry of Natural Resources. 1995. Ontario Community Forest Pilot Project. Lessons Learned 1991-1994. Sault Ste. Marie.

Resource Information Systems Inc. (RISI) and Resource Economics. 1992. Ontario Forest Products and Timber Resource Analysis. Queen's Printer: Toronto.

Wildlands League, Forest Diversity–Community Survival Project. 1995. Ontario's Forest Products Industry–A New Appetite in the Forest. Toronto.

PART THREE:
SCIENTIFIC CONSIDERATIONS

Prince Albert
Model Forest
Association Inc.

MODEL FOREST
NETWORK
RESEAU DE
FORÊTS MODÈLES

An Ecological-Economic Analysis of the Role of Canadian Forests in Mitigating Global Climate Change

Mohammed Dore
Mark Johnston
Harvey Stevens

INTRODUCTION

The forests of Canada comprise a resource-environmental system that not only provides timber and pulpwood, but also offers recreational benefits, performs an important soil stabilization function and plays a vital role in the absorption of atmospheric carbon dioxide, thus mitigating global warming. The purpose of this paper is to estimate the value of the forests in mitigating global warming.

The atmospheric concentration of CO_2 has increased about 25% above pre-industrial levels, partly due to the burning of fossil fuels and partly due to deforestation. If present trends persist, it is ex-

Mohammed Dore is Professor of Economics at Brock University, Ontario, Canada.

Mark Johnston is with the Ontario Ministry of Natural Resources, Ontario, Canada.

Harvey Stevens is Research Associate at Brock University, Canada.

[Haworth co-indexing entry note]: "An Ecological-Economic Analysis of the Role of Canadian Forests in Mitigating Global Climate Change." Dore, Mohammed, Mark Johnston, and Harvey Stevens. Co-published simultaneously in *Journal of Sustainable Forestry* (Food Products Press, an imprint of The Haworth Press, Inc.) Vol. 5, No. 1/2, 1997, pp. 199-206; and: *Sustainable Forests: Global Challenges and Local Solutions* (ed: O. Thomas Bouman, and David G. Brand) Food Products Press, an imprint of The Haworth Press, Inc., 1997, pp. 199-206. Single or multiple copies of this article are available for a fee from The Haworth Document Delivery Service [1-800-342-9678, 9:00 a.m. - 5:00 p.m. (EST). E-mail address: getinfo@haworth.com].

199

pected that temperatures will increase as a result of the "greenhouse effect" an estimated 0.3 degrees C per decade, potentially affecting human health, agricultural productivity, and sea levels. Stabilizing the concentration of CO_2 at present levels would require an estimated 60 percent reduction in CO_2 emissions (Faeth et al. 1992). If "global warming" is a problem, then forests and forest soils may be part of the solution, as they provide society with a valuable service. Effective sustainable management of the forest resource over time requires that analysts take into account the ecological services of the forest resource-environmental system in making the harvesting decision. *This implies combining economics with forest ecology.*

This paper is part of a larger project funded by the Social Sciences and Humanities Research Council of Canada. The aim of the project is the development of a dynamic optimization model of forest harvesting in which the decision as to when to cut, how much to cut and when to stop are all endogenously determined. The policy importance of such a model suggests that forests should be treated as *social capital assets*, with known ecological functions that are non-marketed. Like all ecological services, the carbon uptake service provided by forests is of the non-marketed type and thus falls within the purview of environmental economics that specializes in imputing values for goods and services not traded in markets. The task of this paper is to estimate the social value of one of the services provided by the forests, namely the carbon uptake service provided by Canadian forests.

As a participant at the Toronto Conference on Atmospheric Change, held June 27-30 1988, Canada is committed, in principle, to a reduction in carbon dioxide emissions of 20 percent of 1988 levels by 2005. If the forests of Canada were not there, the carbon that they process would have to be cleaned up from the atmosphere, which would carry a certain cost. This annual cost represents a portion of the capital value of the forests. In conventional optimal harvesting models, the capital value of forests is reflected only in stumpage and amenity values (Conrad and Ludwig unpublished). By contrast, our model of the optimal harvesting decision would incorporate the value of the carbon uptake service.

METHODOLOGY

Carbon is released into the atmosphere through natural processes, decomposition, deforestation, and fossil fuel burning. From the social viewpoint, the forest acts as a free factory, accepting a certain amount (megatonnes) of carbon per annum for "processing" via photosynthesis, which is sequestered in the forest biomass and forest soils.

The rate at which trees as "machines" can successfully process carbon depends on their "technological efficiency." The rate at which trees fix carbon depends on the biomass growth rate which in turn depends on the CO_2 concentration in the atmosphere. Short-term experiments under controlled conditions in glasshouses show that increased levels of CO_2 in the atmosphere promote increased rates of photosynthesis and growth in most plants. Were this to translate to natural ecosystems, the result would be a significant negative feedback on global warming. However, because of lack of data, this feedback effect has been omitted from the analysis that follows.

If there existed some technology with the ability to remove carbon dioxide from the atmosphere, then proceeding with valuation would be straightforward. The cost of operating this technology would serve as a proxy for the value of the service provided by the forest. Engineering approaches to enhance the ecosystem's carbon dioxide assimilative capacity, such as activating carbon-eating organisms in the oceans or injecting sulphur dioxide into the stratosphere to absorb visible sunlight have been proposed (Baum 1990 and Schneider 1990) but for practical purposes, such technologies do not exist. While such technologies may be developed in the future, they are relatively expensive. The preferred cost effective method of mitigating global warming is still reducing emissions, and conserving forests.

Conventional economic estimates of the effects of emission reduction in terms of lost output are typically carried out using production functions (Cline 1992). However, the results would be highly sensitive to the choice of the functional form of the production function. Similarly, Pearce (1991) estimates the value of forests by assigning a "credit" for damage avoidance through conservation of the forests or a "debit" for alternative land use. Thus, Pearce measures "opportunity cost" in terms of damage repair. However, estimates of damage repair are highly subjective and controversial. In our work, we avoid both of these approaches.

Our approach to valuation is in terms of foregone ecological services. It is the services that we would lose if the forests were not there. This is more direct and closer to the classical meaning of the concept of opportunity cost. In this way we also avoid the problems encountered in attempting to simulate a non-existent market, which is a requirement in the Contingent Valuation Method.

We interpret the value of the carbon uptake service as the value added in industry that generates the same amount of carbon dioxide that the forest is capable of processing, where value added is the value of the economy's output minus the value of the inputs. Thus, a portion of the cost to society of harvesting forests is represented by foregone industrial output.

The net flux of carbon (C) between the forest ecosystem and the atmosphere determines whether forests are part of the problem or part of the solution with regard to changes in the concentration of CO_2 in the atmosphere. Estimates for Canada of the carbon uptake by forest biomass and forest soils are revealed in *The Carbon Budget of the Canadian Forest Sector: Phase I* (Kurz et al. 1992). According to this study, in the reference year 1986, the C pools in Canadian forest ecosystems increased by 29.5 Mt, which represents the sum of the biomass and soil C pool changes. We denote the level of carbon uptake by the forest biomass and forest soils as Q.

Marland et al. (1994, pp. 554-556) provide a time series on Canada's carbon dioxide emissions from 1950. They estimate the level of carbon dioxide emissions at 106.8 Mt C in 1986. Thus, in its role as factory, the Canadian forests process 29.5 Mt C of the 106.8 Mt C released through industrial activity (see Figure 1). The value of manufacturing output in 1986 is estimated at $253,410.6 million and the cost of inputs at $153,335.7 million for a value added of $100,074.9 million (Statistics Canada). Given this information, we proceed with the valuation exercise as follows:

If 106.8 Mt C generated corresponds to industrial output of $253,410.6 million, then 1 Mt C corresponds to Y dollars.

FIGURE 1. The forest as nature's factory (Canada, 1986)

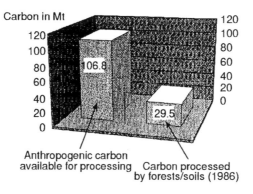

Carbon in Mt

Anthropogenic carbon
available for processing

Carbon processed
by forests/soils (1986)

Therefore, $Y = \$253,410.6$ million/106.8 Mt C
$= \$2,372.76$ million/Mt C.

Recall that the level of carbon uptake is represented by Q. Thus, the value of industrial output corresponding to the amount of carbon that the forests process is QY dollars:

$$QY = \$29.5 \text{ Mt C} \times \$2,372.76 \text{ million/Mt C}$$
$$= \$69,996.42 \text{ million.}$$

If the cost of input per unit of output is Z dollars, then the cost of all inputs is ZQY dollars.

Cost of input per unit of output, $Z = \$153,335.7$ million/253,410.6 million.

Therefore, $Z = 0.6051$

Thus, the cost of all inputs, $ZQY = 0.6051 \times \$69,996.42$ million.
$= \$42,354.83$ million.

The value of the carbon uptake service V is calculated based on the equation:

$$V = QY (1 - Z)$$

where:

Q represents carbon uptake by the forest biomass and forest soils,

Y is the ratio of the value of output of the manufacturing sector to the level of emissions,

Z is the ratio of the cost of inputs to the value of outputs.

RESULTS

Substituting the values determined above into the valuation equation:

$$V = QY - ZQY$$
$$= QY(1 - Z) \text{ million dollars in the year 1986}$$
$$= \$69{,}996.42 \times (1 - 0.6051) \text{ million}$$
$$= \$27{,}641.58 \text{ million.}$$

We determine the value of the carbon uptake service provided by Canadian forests in the year 1986 to be approximately *$27 billion*.

Given that GDP in Canada in 1986 was approximately $506 billion, the value added as a result of the ecological service of carbon uptake provided by the forests in 1986 was about 5.5% of GDP (see Figure 2). In addition, the value added by all manufacturing in 1986 was approximately $100 billion with value added in the forest products industry at $6.5 billion. Clearly, the value of the carbon uptake service provided by forests is significant and should be taken into account in the determination of the harvesting decision.

At present there does not exist a time series on carbon uptake. However, dividing both sides of the equation $V=QY(1 - Z)$ by Q we derive the value of the carbon uptake service expressed as millions of dollars *per Mt of carbon uptake* or $V/Q=Y(1 - Z)$. For the year 1986, the value of the carbon uptake service is thus $937 million dollars per Mt of carbon uptake. In this way we can derive

FIGURE 2. Value of carbon uptake relative to GDP (Canada, 1986)

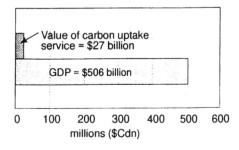

an estimate, per Mt of carbon, of the ecologically acceptable level of industrial output that would not contribute to global warming, provided the forests exist in their present size. The ecologically acceptable level of output depends on the estimate of carbon uptake provided by forestry ecologists. In further research we hope to fit a seasonal ARIMA model to the time series:

$$(V/Q)_t = (1 - Z)Y_t$$

in order that we may determine the value of the carbon uptake service with 95 per cent confidence. (This research work is in progress.)

If we are to incorporate the value of non-marketed services in a model of the optimal harvesting decision, then there is a clear research need for a time series of data on carbon uptake.

SUMMARY

Given the character of the manufacturing (emissions levels and composition of output), we find that approximately $28 billion worth of manufacturing value added is made possible by the carbon uptake service provided by Canadian forests in 1986, with no corresponding increase in emissions. Thus, forests play the role of a "free factory." The above figure may also be expressed per Mt of carbon uptake, as approximately $937 million. Then, based on estimates of carbon uptake provided by forestry ecologists, we impute a value of the non-market service of carbon uptake that forests provide. The calculation of such a value contributes to the determination of the value of forests as social capital. This value must then be incorporated in the model that determines the optimal harvesting decision. Without this value, the social cost of forests is seriously underestimated, which would lead to over-exploitation of forest resources.

REFERENCES

Baum R. (1990) "Adding Iron to Ocean Makes Waves as Way to Cut Greenhouse CO_2," Chemical and Engineering News (July 2): 21-24 cited in van Kooten, G. G., et al. (1992) "Potential to Sequester Carbon in Canadian Forests: Some Economic Considerations." Canadian Public Policy. XVIII:2:127-38.

Cline, William R. Global Warming: The Economic Stakes. Institute for International Economics: (Washington, 1992).

Conrad, Jon and Don Ludwig. "Forest Land Policy: The Optimal Stock of Old-Growth Forest," unpublished.

Dore, Mohammed H. I. (1996). The Problem of Valuation in Neoclassical Environmental Economics, Environmental Ethics, Spring 1996, Vol. 18, pp. 65-71.

Faeth, Paul, Cheryl Cort and Robert Livernash. Evaluating the Carbon Sequestration Benefits of Forestry Projects in Developing Countries. World Resoures Institute, February, 1992.

Kurz, W. A., M. J. Apps, T. M. Webb, P. J. McNamee. (1992). The Carbon Budget of the Canadian Forest Sector: Phase I. ENFOR Information Report NOR-X-326 (Edmonton: Forestry Canada, Northwest Region).

Marland, G., R. J. Andres, and T. A. Boden. (1994). "Global, regional and national CO_2 emissions." pp. 505-584. In: T. A. Boden, D. P. Kaiser, R. J. Sepanski and F. W. Stoss (eds). Trends '93: A Compendium of Data on Global Climate Change. ORNL/CDIAC-65. Carbon Dioxide Information Analysis Center, Oak Ridge National Laboratory, Oak Ridge, Tenn., U.S.A.

Pearce, David. An Economic Approach to Saving the Tropical Forests. Reprinted in Economic Policy Towards the Environment. Dieter Helm, ed., Blackwell: (Cambridge, Mass., 1991).

Schneider, S. (1990) The Greenhouse Effect: Science and Policy, Science, 243:777-81.

Selected Forestry Statistics, Canada, 1988. Information Report E-X-41. Economics Branch, Forestry Canada, June, 1989.

Statistics Canada. Canada Yearbook, 1994.

Value of Wilderness Protection in Saskatchewan: A Case Study of Existence Values

S. N. Kulshreshtha
K. G. Loewen

ABSTRACT. Preserved wilderness in Saskatchewan has value to both the users and to non-users. To determine such values, a study of Saskatchewan residents' level of willingness-to-pay (WTP) for wilderness preservation was carried out. Results suggest that 58% of sample non-aboriginal and only 7% of aboriginal respondents were satisfied with the current level of wilderness and wildlife protection in the province. Their reasons for preserving wilderness areas varied but protection of water quality, air quality, and wildlife habitats topped the list. The estimated annual WTP to ensure current wilderness protection prevails was approximately $61 per household for the non-aboriginal sample, and $80 for the aboriginal sample, which amounts to a value of $100 per hectare. *[Article copies available for a fee from The Haworth Document Delivery Service: 1-800-342-9678. E-mail address: getinfo@haworth.com]*

S. N. Kulshreshtha is Professor and K. G. Loewen is Professional Research Associate, in the Department of Agricultural Economics, University of Saskatchewan, Canada.

Financial assistance received from the Prince Albert Model Forest Association is gratefully acknowledged.

[Haworth co-indexing entry note]: "Value of Wilderness Protection in Saskatchewan: A Case Study of Existence Values." Kulshreshtha, S. N., and K. G. Loewen. Co-published simultaneously in *Journal of Sustainable Forestry* (Food Products Press, an imprint of The Haworth Press, Inc.) Vol. 5, No. 1/2, 1997, pp. 207-216; and: *Sustainable Forests: Global Challenges and Local Solutions* (ed: O. Thomas Bouman, and David G. Brand) Food Products Press, an imprint of The Haworth Press, Inc., 1997, pp. 207-216. Single or multiple copies of this article are available for a fee from The Haworth Document Delivery Service [1-800-342-9678, 9:00 a.m. - 5:00 p.m. (EST). E-mail address: getinfo@haworth.com].

NEED FOR THE STUDY

Wilderness is an important resource in Saskatchewan. Wild spaces, according to the World Wildlife Fund Canada (1993, p. 28), everywhere are shrinking or vanishing altogether before relentless human pressure. At the present time, the province has a total of 4.2 million hectares of protected area under various jurisdictions. A larger majority of this land is under provincial jurisdiction, and includes provincial parks, community pastures, wildlife refuges, and wildlife habitat protection lands, among others. Even in Saskatchewan, with a very small population base, and relatively lower rates of urban sprawl, protection of wild lands has not been considered adequate.[1]

From the perspective of forest land management, preservation of wilderness is among the important activities to consider. This requires knowledge of non-use values of such lands by Saskatchewan residents. Ignoring non-use values in natural resource policy making could lead to misallocation of resources. Furthermore, since the issue of wilderness protection is attracting public attention and has already undergone some public scrutiny, it was felt necessary to seek public input for the planning process and for wilderness management in the province. This input may take the form of how the public values protection of the wilderness in the province. An important aspect of forest management is the cultural perspective. Since the Prince Albert Model Forest (PAMF) and the surrounding region houses aboriginal communities, it is important to incorporate their views about wilderness preservation as well.

OBJECTIVES OF THE STUDY

The primary objectives of this study are: (1) to report on solicited opinion of Saskatchewan residents about nature of wilderness preservation in the province; (2) to develop and apply a procedure for measuring the benefits of wilderness resources as perceived by the residents of Saskatchewan; and (3) to compare and contrast the differences between aboriginal and non-aboriginal residents' value of wilderness in the province.

ECONOMIC VALUE OF WILDERNESS PROTECTION

Wilderness provides numerous benefits to society, many of which may not be perceived to be directly related to commercial economic activities. Some of these benefits are derived through provision of environmental amenities, which affects the long run capacity for economic activities, and efficiency of resource use. Benefits of wilderness may accrue to society through actual use of the wild areas. However, studies have suggested that a significant part of the value of wilderness may exist for reasons other than for its use. The total economic value of a wilderness area, thus, can be delineated to be a sum of two types of values: Use values or non-use or passive values. The latter types of values are identified for those people who may associate a value but do not belong to a particular user group. Thus, non-use values result from the sense of satisfaction that the nature resource will be preserved and that benefits from it would continue to flow now and in the future.

Establishing a price or value of wilderness is difficult because there is no organized market where these types of goods are traded. Thus, it is customary to estimate the benefits of such experiences, which would include those originating through use as well as non-use, by using non-market methods of valuation. Although two types of methods are available for estimating such values (the Travel Cost Method, and the Contingent Valuation Method—CVM), the latter approach is more appropriate for estimating both use and non-use values.

The CVM approach provides the respondents with a hypothetical situation relevant to the valuation exercise, which includes the description of a proposed program. The respondents are then confronted with a question or questions that provide the economic sacrifice they are willing to make for the proposed program. Although the CVM has been a subject of great controversy among economists and others, the NOAA Panel has recently concluded that "it can produce estimates reliable enough to be the starting point of a judicial process of damage assessment, including lost passive values" (Federal Register 1993, p. 4610).

STUDY METHODOLOGY

In this study, the CVM was used to estimate preservation values of wilderness in Saskatchewan. The approach taken was that of an open-ended contingency valuation to solicit respondents' WTP for preservation. This method involved a contribution into a special fund designed specifically for the purpose of preserving and maintaining wilderness. This relatively neutral method of payment was chosen because an entrance fee or tax would likely result in numerous protest responses.

A questionnaire was developed to elicit the respondents' bids. It focused on wilderness and recreation activities by residents, their opinions on the current preservation programs, as well as on their demographic, economic, attitudinal and behavioral information. The data collection was divided into two sets, one each for the aboriginal and the non-aboriginal households.

For the aboriginal sample, a sample of 30 households living in the city of Prince Albert was selected. Since no reliable list of names and addresses of such families existed, an interviewer was hired and families were contacted at random in the predominantly aboriginal sections of the city. All information was collected through face-to-face interviews.

For the non-aboriginal sample, the province was divided into four zones and a stratified random sample of 850 was selected initially. For some regions, response rates were poor. For this reason, a second mailing of 400 surveys was carried out. Both rounds of mailings resulted in a total of 238 surveys, of which 193 were useable, resulting in a useable return rate of 16%.

RESULTS

Characteristics of Sample Respondents

The average non-aboriginal study respondent was 45 years of age, had a household size under three persons and a household income of slightly more than $46,000. The average education level of these respondents was near 13 years. The mean number of children under 16 years of age per household was under one. Nearly a quarter of those surveyed were retired. In contrast, the average aboriginal family was larger in size, with an average of 3.83 members per family. The average respondent tended to be younger (only 42.4 years old), with a relatively younger family (as denoted by an average of 2.13 members below the age of 16 years). The average family income was lower, only $17,154 per annum.

Motivations for Preservation

The motivation for preservation differed between the non-aboriginal and aboriginal households. Results are shown in Table 1. All motivations had a mean score of three or more, indicating they were positive motivations towards preservation of wilderness. The non-aboriginal respondents indicated that the most significant reasons for preserving wilderness were for the purposes of protecting water and air quality, rare and endangered species and wildlife habitats. These were closely followed by preserving it for the future generations–a concept very close to the bequest value. Providing spiritual inspiration and tourism dollars were the least important of the reasons to preserve; however, they were definitely not unimportant reasons. The average aboriginal respondent had significantly higher scores for the preservation reasons than the average non-aboriginal respondent. Although reasons for preservation were very similar between the two groups, the aboriginal households are more concerned with the health of the environment for current and future generations, while the non-aboriginal respondent placed a somewhat greater emphasis on commercial aspects such as tourism.

Satisfaction with Current Preservation of Wilderness

Of the non-aboriginal respondents surveyed, 58% indicated that they were satisfied with the current level of wilderness and wildlife protection in the province, while the remainder (42%) wanted more protection. In contrast, only 7% of the aboriginal respondents were satisfied with current wilderness protection in Saskatchewan. A similar study in British Columbia (Reid and Stone 1994), indicated that only 37% of B.C. residents were satisfied with current wilderness protection.

TABLE 1. Importance of motivations for preservation of wilderness by aboriginal and Saskatchewan households*

Reason	Aboriginal (27-29 respondents)		Non-Aboriginal (213 respondents)	
	Mean Score	Std. Dev.	Mean Score	Std. Dev.
Water Quality	5.00	0.00	4.62	0.75
Future Generations	5.00	0.00	4.54	0.83
Knowing Wilderness Exists	4.97	0.19	4.29	1.05
Rare/Endangered Species	4.97	0.19	4.56	0.78
Air Quality	4.93	0.37	4.58	0.83
Wildlife Habitat	4.90	0.41	4.56	0.82
Scenic Beauty	4.90	0.31	4.43	0.79
Spiritual Inspiration	4.78	0.51	3.21	1.31
Unique Environments	4.61	0.88	4.34	0.91
Option to Visit in Future	4.61	0.99	4.34	0.93
Recreational Opportunities	4.21	1.07	4.35	0.84
Educational and Scientific Study	3.89	1.40	4.15	0.99
Revenue from Tourism	3.13	1.60	3.60	1.16

*A scale of 1 (not at all important) to 5 (extremely important) is used.
A range is reported in the number of respondents because some respondents skipped some of the questions.

Estimation of Relationship Between WTP and Respondents' Characteristics

The sample households were asked to express their WTP for preservation of wilderness. The question was presented in three parts, which asked for their WTP for current levels of preservation, as well as for an increase from 5% to 7.5% and from 5% to 10%. The range of the WTP was non-negative in this study, but contained several zero bids. The average WTP for the sample non-aboriginal respondents was $60.89 for current preservation, where as the aboriginal respondents' offer was $80, some $19 more. This value is very similar to that found by other studies. For example, Walsh, Gillman and Loomis (1982) reported wilderness values ranging from $26-$85 per Colorado household in 1980. Reid and Stone (1994) reported a mean value of $119 per household for the doubling of preserved wilderness in British Columbia.

Distribution of Estimated Willingness to Pay

The willingness-to-pay (WTP) for wilderness preservation in Saskatchewan varied from zero level to a high positive value. A typical range for both the samples was between 0 to $1000 for current level of preservation. A comparison of the aboriginal and non-aboriginal respondents' WTP suggests that aboriginal people are willing to pay higher for preservation than their non-aboriginal counterparts (Figure 1). For example, of the total of 193 non-aboriginal respondents, 44% were not willing to pay anything for

FIGURE 1. Willingness-to-pay for current preservation; aboriginal and non-aboriginal respondents, Saskatchewan, 1994

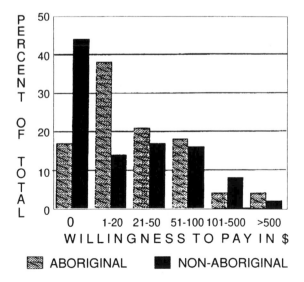

current wilderness preservation, 47% were willing to pay between $1-$100 and the remaining 9% were willing to pay more than $100. Further analysis of the sample also suggested that membership in wilderness groups affected the WTP for current level of preservation. Respondents associated with such groups, for the last five years were willing to pay nearly twice as much for current levels of preservation as non-members.

Total Economic Value of Wilderness Preservation

The willingness-to-pay for preservation of wilderness was used to estimate its total economic value. For the case presented in this study, many respondents have used the wilderness for a variety of purposes. To obtain distribution of this valuation among various motives for preservation, further elicitation was needed. Respondents were asked to allocate their contribution amount among four options, namely, recreation, bequest, existence and option values (Table 2). Approximately 69.4% of the contributed amount was allocated to existence and bequest values. The remaining 30.6% was for current recreation or for the option to use it in the future.

If we assume that existence and bequest values are close to non-use values, and given that these uses constitute roughly 69.4% of the total, one can estimate the non-use value of wilderness in Saskatchewan. For the non-aboriginal households, using the stated WTP and a total of 336,000 families (as per 1991 census), the non-use value of preservation is estimated at $14.2 million. Similar-

TABLE 2. Recreation and preservation allocation shares for non-aboriginal sample respondents, Saskatchewan and British Columbia residents

Value	Percent of Total	
	Saskatchewan	British Columbia
Existence Value	37.5	37.8
Bequest Value	31.9	38.7
Option Value	15.1	13.2
Recreation Value	15.5	10.3

ly for the aboriginal households (number of households estimated at 26,000 for 1991) a total value of $2.2 million is obtained. Using a 3% rate of discount, the total value of $16.4 million per year yields a value of $422 million for preserved wilderness, or about $100 per hectare.

Economic Value of Increased Preservation

Respondents were also asked as to their WTP for a doubling of the protected wilderness area. Most of the non-aboriginal respondents responded positively to such an increase. Their WTP for such activities either remained the same or increased from the previous level. Overall, 71% of respondents showed no change, whereas 12% showed an increase. The average level of WTP for doubled wilderness area was estimated at $88.82 per household. This yields a marginal value of $27.93 per household for another 4.035 million hectare of wilderness areas. Total economic values of adding another 4.025 million hectare of wilderness is estimated at $10.11 million per year, which yields a value of $64.63 per hectare of additional preserved lands. This value reflects the marginal economic value of preserving wilderness in the province of Saskatchewan.

NOTE

1. For example, the Saskatchewan Round Table on Environment and Economy (Undated) has recommended for an Endangered Species and Habitat Act, and under that act, has suggested for further designation of exceptional and vulnerable sites in the province. Although information on change in the area under protected lands is not available, the World Wildlife Fund Canada has evaluated the provincial wilderness protection program unsatisfactory (Star-Phoenix, April, 1995).

REFERENCES

Federal Register (1993). "Appendix I–Report of the NOAA Panel Contingent Valuation." 58(10):4602-4614.

Reid, R. and M. Stone (1994). Economic Value of Wilderness Protection in British Columbia. British Columbia Ministry of Environment, Lands and Parks; and British Columbia Ministry of Forests.

Saskatchewan Round Table on Environment and Economy (Undated). Conservation Strategy for Sustainable Development in Saskatchewan. Regina.

Walsh, R., R. Gillman, and J. Loomis (1982). Wilderness Resource Economics: Recreation Use and Preservation Values, Denver, Colorado: America Wilderness Alliance.

World Wildlife Fund Canada (1993). Protected Areas and Aboriginal Interests in Canada. WWF Canada Discussion Paper, Ottawa.

Sustaining Tropical Forest Biodiversity

P. D. Khasa
B. P. Dancik

ABSTRACT. Tropical forests are very rich in biological diversity and form an important economic and ecological resource. This biodiversity is of great value for communities living in or near these forests as a ready source of subsistence and cash income, and for the world at large as a source of tropical timber and non-timber products and a repository of genetic and chemical information. However, this biological complexity is diminishing rapidly. We analyze the main human (anthropogenic) actions causing loss of tropical forest biodiversity along with the strategies of management for forest biodiversity. Not only is the biophysical component important in management for biodiversity, but the action participation and support of local people, the national government and international cooperation as a whole, are essential for an effective and sustainable development of tropical forests. This integrated development strategy ensures that socio-economic and environmental benefits are provided for present

---

P. D. Khasa and B. P. Dancik are affiliated with the Department of Renewable Resources, University of Alberta, Edmonton, Alberta, Canada.

Address correspondence to: P. D. Khasa.

The authors are grateful to Dr. O. P. Rajora (Department of Renewable Resources, University of Alberta) for his valuable comments on previous drafts of this manuscript. Support provided by Natural Sciences and Engineering Research Council of Canada Grant A0342 to B. P. Dancik is gratefully acknowledged. The authors also thank the Biodiversity Convention Office in Ottawa, for access to documents valuable in making this review.

An extensive version of this paper has been published in the *Journal of Sustainable Forestry*, Volume 4(1/2), 1997, pp. 1-31.

and future generations. *[Article copies available for a fee from The Haworth Document Delivery Service: 1-800-342-9678. E-mail address: getinfo@haworth.com]*

INTRODUCTION

At the end of 1990, tropical forests covered some 1,756.3 million ha and the estimated area of forest plantations was 30.7 million ha with an increase of about 1.8 million ha per year (Singh 1993). Tropical forests covered 918.1 million ha (52%) in Latin America and the Caribbean, 527.6 million ha (30%) in Africa, 310.6 million ha (18%) in Asia and the Pacific. Table 1 presents a classification of tropical forests, their population development, and annual deforestation as covered by the Food and Agriculture Organization's (FAO's) 1990 tropical forest resources assessment (Singh 1993). The FAO estimated that about 15.4 million ha of tropical forests disappear annually worldwide, with 4.1 million ha in Africa, 3.9 million ha in Asia and the Pacific, and 7.4 million ha in Latin America and the Caribbean (Singh 1993). The overall average rate of deforestation is 0.8% per year (Table 1).

Tropical forest biomes are extremely diverse, including deserts and semi-arid savannas, dry forests (Muhulu), open woodlands (Miombo), rain forests, mountain forests and coastal mangroves (Trochain 1980). In this report, we define biodiversity (or biological diversity) as 'the variety and variability among living organisms, and the ecological complexes in which they occur,' encompassing different genes, ecotypes, species, ecosystems, landscapes, and their relative abundance (US Congress OTA 1992). The rain forests have the highest biodiversity, containing at least 50% of the world's species of Monera, Protista, Fungi, Plantae and Animalia (Myers 1989; Wilson 1989). Some authors, like Kimmins (1991), have suggested, however, that the biodiversity of microorganisms in the generally poorer tropical forest soils may support fewer species than temperate and boreal forest soils. On the other hand, biodiversity changes over time and space; this should be taken into account before making any comparisons. Therefore, Khasa and Dancik (1997) have suggested that both epigeous and hypogenous biodiversity would be much higher in tropical rain forests than in any other terrestrial ecosystems.

TABLE 1. Types of tropical forests, their population development and annual deforestation as covered by the FAO's 1990 tropical forest resources assessment project (compiled from Singh 1993)

Biomes	Tropical forest area 1990			Population		Annually deforested area (1981-1990)	
	million ha	%	Forested area (%)	Density (inhab./km^2)	Growth (%)	million ha	%
Rain forest	718.3	41	76	41	2.5	4.6	0.6
Moist deciduous forest	587.3	33	46	55	2.7	6.1	0.9
Dry deciduous forest	178.6	10	25	106	2.4	1.8	0.9
Very dry zone	59.7	3	11	24	3.2	0.3	0.5
Total lowland types	1543.9	88	44	57	2.5	12.8	0.8
Moist upland forests	178.1	10	34	52	2.7	2.2	1.1
Dry upland forests	26.2	2	15	70	3.2	0.3	1.1
Total upland types	204.3	11	29	56	2.9	2.5	1.1
Non-forest (desert and alpine) zones	8.1	<1	1	15	3.5	0.1	0.9
Total tropical forests	1756.3	100	37	52	2.7	15.4	0.8

We do not know exactly how many species exist on our planet but based on recent work in tropical rain forests, the total number of species is estimated as high as 100 million, of which only about 1.7 million living species of all kinds of organisms have been described (WCMC 1992). However, species are eliminated at a rate of 10,000 per year in all tropical forests before even being discovered (Myers 1989; Wilson 1989). Because of the world's fast-diminishing tropical forests, a number of actions have been taken by the international community to save the remaining tropical forests such as the Tropical Forestry Action Programme and proposals of the United Nations Conference on Environment and Development's Agenda 21 (Khasa and Dancik 1997). In this report, we analyze the values of biodiversity, the main human actions causing loss of tropical forest biodiversity along with the technologies to sustain tropical forest biodiversity.

THE VALUE OF TROPICAL FOREST BIODIVERSITY

Tropical forest resources provide many renewable goods and services essential for human welfare (Holdgate 1993; Khasa et al. 1995). They are a ready source of subsistence and cash income for billions of people living in and near the forests by providing agricultural products, fuelwood and charcoal, materials for rural pit-sawing and forest industries, materials for rural construction and fencing, and other non-wood forest products such as forage, fruits, oil-producing seeds, honey, meat, fish, water, beverages, mushrooms, insects, larvae, cosmetics, rubber, gums, resins, waxes, tannins, medicines, and opportunities for ecotourism. The benefits from these forests are also extended to the world at large as a source of timber and non-timber products and a source of genetic and chemical information (Khasa and Dancik 1997). Recent statistics on the production and export of forest products are available from the *Yearbook of Forest Products* and those of crops from the *Yearbook Production* and the *Quarterly Bulletin of Statistics*, published by the FAO. However, statistics on the production and export of non-fibre forest products are scanty or non-existent even though the values of non-fibre products are significantly higher than returns from alternative land uses (Khasa and Dancik 1997). One example is the Madagascar rosy periwinkle (*Catharanthus roseus*), which is the source of vinblastine and vincristine used in the treatment of Hodgkin's disease and childhood leukaemia, and worth at least US $160 million in sales every year (Burton et al. 1992). A number of tropical plant species are also a source of biopesticides and intoxicants (Mandava 1985; Stoll 1988).

Biodiversity is ecologically important for the regeneration and maintenance of forest ecosystems. Insects and other animals play a very important role as pollinators of plants to maintain ecosystems. For example, insect-pollination of oil-palm trees in Malaysia yields annual savings of about US $140 million from hand-pollination (Pimentel et al. 1992). The majority of tropical tree species are zoochorous in noninundated forests (Pendje and Baya 1992), and hydrochorous or ichthyochorous in flooded forests (Goulding 1993; Kubitzki and Ziburski 1994). The role of soil biota as components of sustainable agroecosystems and forest ecosystems or for use as biofertilisers in agriculture, horticulture, and forestry is well documented (Hendrix et al. 1990; WCMC 1992; Reddy et al. 1997). Finally, forests play an important role in watershed regulation and stabilization of soils in erosion-prone areas, in protection of estuaries, and as a carbon sink for climate regulation at the global level (WCMC 1992).

Many domesticated plant and animal species have wild relatives that are sources of genetic material for the improvement of productivity, nutritional value, and tolerance to insects, diseases, and environmental stresses (Burton et al. 1992). New varieties are now being created by traditional selective breeding or by genetic engineering for production of transgenic organisms upon which evolutionary processes can work. For example, an increased content of soluble solids in commercially-grown tomatoes has been obtained through a program of genetic crosses between a single wild tomato species (*Lycopersicon chmielewskii*) collected in Peru and commercial varieties, contributing US $8 million a year to the American tomato-processing industry (Burton et al. 1992). In a number of tree species (poplars, pines and eucalyptus), conventional breeding technology or genetic transformation has led to the recombination of genes with the ultimate aim of making trees resistant to pathogens, tolerant to herbicides, and for introducing other desired traits.

Diffuse (non-utilitarian) values of biodiversity play a role as a provider of social, cultural, scientific, and recreational benefits, human health and equilibrium (Bouman and Brand 1997). For some culture, aesthetic, spiritual, and ethical motivation and self-interest are playing a major role in promoting public and private programs to conserve particular species or habitats. Such justification, based on aesthetic, spiritual, and ethical arguments, are not directly quantifiable in dollars. However, Panayotou, DeShazo and Vincent at the Harvard Institute for International Development, Cambridge, MA have been trying to quantify aesthetic value (personal communication).

CAUSES OF LOSS
OF TROPICAL FOREST BIODIVERSITY

The main human causes affecting the loss of biodiversity in tropical forests have been reported by many authors (e.g., Steinlin 1994; Khasa et al. 1995; Khasa and Dancik 1997). Slash-and-burn agriculture is widespread and remains the primary cause of loss of tropical forest biodiversity. Most of the trees felled in this type of agriculture are not even used for lumber, but are burned *in situ* (Panayotou and Ashton 1992). Income generating uses of the non-fibre products for local, regional, national or international markets can also lead to unsustained extraction (over-exploitation) of animal and plant species and loss of biodiversity (e.g., rhinoceros and elephant horns, and medicinal plants).

Use of wood energy is another important cause of tropical deforestation. Of the 3,400 million m^3 of roundwood felled for human purposes in 1991, 53% had been used as a source of energy in the form of fuel wood and charcoal, 89% of it in developing countries (Steinlin 1994). The total fuelwood and charcoal consumption in developing countries was 1,600 million m^3 in 1991 and is growing more or less parallel to population growth (Steinlin 1994). Wood energy not only serves domestic purposes but also local small industries.

Depending on its intensity, the setting of fires during the dry season for hunting, cattle grazing, and shifting agriculture is another agent of loss of both epigeous and hypogenous biodiversity; it also contributes to soil erosion and atmospheric warming (greenhouse effect). Steinlin (1994) estimated that about 3,400 million m^3 had been destroyed worldwide by forest fires. Open woodlands, savannas, and the dry forests are fire-prone ecosystems during the dry season.

The forest industry is still in its infancy in many tropical countries and is not the main cause explaining the decline of tropical forests. Rural production from pit sawing can amount to a sizeable quantity in many countries, especially with the introduction of portable sawmills. Resource mining and highly selective logging systems generally lead to genetic erosion and extinction of certain alleles. Provenance and progeny tests have generally shown that slowly growing and effective trees resulting from dysgenic selection are genetically inferior (Zobel and Talbert 1984). Ground-skidding with crawler tractors also leads to soil disturbance and mortality of seedlings, saplings and poles. Road building without low-impact in mind causes edge effects and fragmentation, among other negative effects. Moreover, forest road building facilitates slash-and-burn cultivators to gain access to the forests and clear patches for temporary agriculture.

Finally, biodiversity in entire ecosystems can be threatened by acts of war and ecoterrorism. In many parts of tropical nations, proscribed pesticides such as DDT or fish intoxicants are still used, contributing to loss of biodiversity. Mining, oil and gas exploration and industrial development without low-impact in mind cause loss of biodiversity. Natural catastrophes (hurricanes, flooding of upland forests, volcanoes, earthquakes, gas emanation) with the interference of humans which has upset the ecosystem balance, constitute a greater threat to ecosystems, now than before.

MEASURES OF BIODIVERSITY

There is little scientific understanding of the composition, structure, and management strategies for heterogeneous tropical forest biodiversity and for rational use based on the principle of sustained yield and assured regeneration. Measures and monitoring are a prerequisite to management for biodiversity. They only become possible when some quantitative values can be ascribed to them and these values can be compared (WCMC 1992). Genetic diversity refers to the variation within and among conspecific populations and among congeneric species. Species diversity, commonly called biodiversity in general terms refers to the number of species within a site or habitat (alpha diversity). The differences between habitats are referred to as ecosystem diversity (beta diversity), while differences in site diversity over large areas (landscape levels of species diversity) are referred to as gamma diversity and the overall gamma diversity of a biogeographic region as epsilon diversity (Burton et al. 1992).

Reid et al. (1993) developed a minimum set of 22 biodiversity indicators providing information useful for local, national, regional, and international policy-makers. To deal with the scale-related problem in developing biodiversity indicators, these authors used indicators of both local extirpation and global extinction, and present measures in both absolute numbers and percentages. However, their measures such as species-richness (number of species per area or habitat type) suffer from the lack of evenness or equitability which refers to the degree of relative abundance of each species in an area. Indices including both richness and abundance were recently presented by Khasa and Dancik (1997). We discuss some of the appropriate diversity measures in the following lines (see also Silbaugh and Betters 1997).

Genetic attributes are one candidate for a favoured measure of diversity. All utilitarian and non-utilitarian justifications for conservation can be applied at the level of genetic attributes. For tropical species, there is little knowledge about intra- and interspecific variation, which is of vital importance for adaptation of species to varying environmental conditions (climate, soil, pests and diseases, etc.), and for breeding programmes. Biochemical and molecular genetic markers are not available to measure genetic diversity and to gain information on quantitative characters of economic and ecological importance (Khasa and Dancik 1997). However, heritable phenotypic and particularly morphophysiological attributes most strongly influence the popular perception of biodiversity, because they are more amenable to the senses than the neutral biochemical and molecular genetic markers (Williams et al. 1994).

Vulnerable and rare species might be considered to have greatest priority for conservation efforts in order to prevent extinctions. Species interactions can result in correlated extinction events and this should be considered when a reserve is established to manage the persistence of a subset of the species present. If there are several vulnerable species, the optimal strategy is to minimize the amount of phylogenetic diversity expected to be lost from the set of vulnerable species (Witting and Loeschcke 1995). However, it may not necessarily be the species most at risk of extinction that are the highest priority, but a set of areas with the most diverse combined biota (Williams et al. 1994). The use of endemism for biodiversity conservation assessment assumes either that widespread species will always co-occur with restricted species, and so will be protected in the same areas, or that most widespread species will persist unaided.

Functional diversity among species is expected to have a premium value for preserving the integrity of ecosystems (Williams et al. 1994). Problems with the measurement of functional attributes are (1) that a species may have many different functions, and (2) that these may vary greatly over time and space (Williams et al. 1994). Also, large information demands are needed for the functional approach, although surrogate variables giving some indication of function may be more readily obtained. Plant functional attributes (PFAs) have been developed for conserving and managing the world's vegetation and are currently being tested as indicators of logging histories and as indicators of growth and yield in tropical forests (Gillison and Carpenter 1994).

Introgressive hybridization between closely-related species pairs in the zones of sympatry may play a significant role in the evolution of these species (Williams et al. 1994). Some authors have argued that introgression increases genetic variability through the production of recombinant genotypes, providing populations a means of coping with environmental change or of evolving novel adaptations (Klier et al. 1991). Alternatively, introgression has been viewed as a primarily local, ephemeral phenomenon with little long-term evolutionary significance. Many widespread species-level taxa show more or less discontinuous geographical variation and thus generate ecotypes, i.e., polytypic species.

In order to assess biodiversity of communities, Ojeda et al. (1995) suggest using a multivariate approach—three biodiversity indicators including species richness, geographic rarity or endemism, and taxonomic singularity or distinctiveness. However, the distinctiveness measure is complicated by the general lack of robust phylogenies, although a rough approximation would be the number of species within each genus. Rapid inventory techniques using satellite imagery, aerial reconnaissance, field surveys, and documents from archives and repositories, can also be conducted. The integration of these data within geographical information systems (GIS) to model variations in the spatial distribution of species richness and to predict where hot spots (geographic locations characterized by unusually high species richness, often of endemic species) occur, produces updated and accurate results in a matter of weeks rather than months or years (Aronoff 1989; Veitch et al. 1995; Parresol and McCollum 1997). GIS allow for development of biodiversity indicators that bear upon both political or biological regions and provide immediate information needed for policy and management decisions in short time scale tackling the difficulties of spatio-temporal changes in biodiversity.

STRATEGIES
FOR SUSTAINING TROPICAL FOREST BIODIVERSITY

The Convention on Biodiversity was ratified by over 90% of United Nations member countries at the UNCED (UNCED 1992). The objectives of the Convention are 'the conservation of biodiversity, the sustainable use of its components and the fair and equitable sharing of the benefits arising out of the utilization of genetic resources, including by appropriate access to genetic resources and by appropriate transfer of relevant technology.' Three countries as dissimilar as Canada, Costa Rica, and Indonesia have at present a national action plant for conservation of biodiversity.

The loss of plant, animal, and microbial resources may reduce future options to develop new important products and processes in agriculture, livestock, fisheries, forestry, medicine, and other sectors; and it may hinder the potential of populations and species to respond or adapt to changing environmental conditions and promote ecosystem stability. Thus the goal of biodiversity conservation is to preserve a representative sample of the total existing diversity by optimizing the *in situ* conservation of biodiversity and minimizing the future loss of biodiversity (Caughley 1994; Witting and Loeschcke 1995). Some of the animal, insect or plant life forms, however, carry to humans new viral infectious diseases such as AIDS caused by HIV, haemorrhagic fever caused by Ebola, Marburg and Lassa viruses. Among others, female mosquitoes of the genus *Anopheles* are vectors of the harmful malarial *Plasmodium*, tse-tse flies (*Glossina* spp.) are vectors of the African trypanosomiases, and molluscs (*Biomphalaria glabrata*) transmit schistosomiasis. Managing for these life forms which represent a serious threat to humans is questionable.

Conservation of biodiversity could proceed at four levels: (1) the diverse landscapes and (2) ecosystems are conserved to preserve diversity of organisms and processes operating at these higher levels of organization, (3) conservation of the species diversity within ecosystems, especially rare and endangered species, and (4) conservation of the intra- and interspecific genetic diversity. To achieve improved conservation and sustainable utilization of biological resources, information is needed on ecological diversity as well as basic information on genetic diversity, spatial distribution of this diversity, and ideally the underlying causes of the observed patterns of genetic diversity (Li et al. 1992).

Sustainable use of biodiversity could be achieved through *in* and *ex situ* conservation measures (Table 2). When no information related to genetic diversity is available, therefore, the optimum *in situ* strategy should be to preserve as many individuals from as many sites, and covering as broad a range of environments as possible (Marshall 1990). But strict preservation is extremely difficult to implement in many sites because of the extreme human population pressures for forest resources. Some innovative plans such as the UNESCO Man and the Biosphere (MAB) program's worldwide network of biosphere reserves, taking into account the needs of the local people by incorporating biophysical and socio-economic factors into the management plan, have been developed for protected areas (US Congress OTA 1992; Figure 1). For species or segments of a species gene pool that are endangered or threatened, *ex situ* conservation is needed for each population at each site (Table 2).

Ledig (1988) has discussed the advantages and disadvantages of *ex situ* and *in situ* techniques and presented in the *in situ* strategy as often the less costly. Consequently, the intensification of *in situ* conservation should be encouraged. For most species that are not under immediate threat but that cannot survive in the long term unless protected, a balance between *ex situ* and *in situ* conservation would be appropriate (Marshall 1990). We need to stress that the use and conservation of biodiversity must be dynamic processes, since human needs, resistance to pests, evolution of pathogens, hosts, and other biotic factors cannot be captured in a fixed state (Namkoong 1991).

Shifting agriculture has been identified as the most important

TABLE 2. Examples of management systems to maintain biodiversity (compiled from US Congress OTA 1992)

In situ		Ex situ	
Ecosystem maintenance	Species management	Living collections	Germplasm storage
National parks	Agro-ecosystems	Zoological parks	Seed and pollen banks
Reserve biospheres	Wildlife refuges	Botanical gardens	Semen, ova and embryo banks
Marine sanctuaries	In situ gene banks	Field collections	Microbial culture collections
Resource development planning	Game parks and reserves	Captive breeding programs	Tissue culture collections
			Cryopreservation

Increasing emphasis on natural processes   Increasing human intervention

FIGURE 1. A schematic diagram of a typical biosphere reserve (adapted from US Congress OTA 1992)

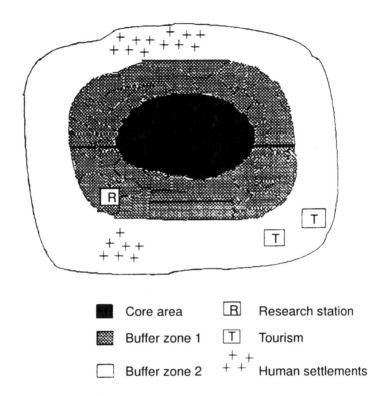

	Core area	R	Research station
	Buffer zone 1	T	Tourism
	Buffer zone 2	+ + + +	Human settlements

agent for loss of forest biodiversity. Thus, improvement of this type of agriculture through more ecological and sustainable agricultural systems is needed (Edwards et al. 1990; Ragland and Lal 1993; Khasa et al. 1995). The maintenance of small livestock at the village level and domestication of some over-exploited animal and plant wild species can better be integrated into agroforestry systems and rural development (Khasa et al. 1995). Indigenous people should be involved in the process and the decisions and should be trained in the improved agricultural systems. A number of successful sustainable projects with addresses of people to write for more information is presented in Katz et al. (1993).

Some countries such as the Peninsular Malaysia, Indonesia, Trinidad and Tobago, Costa Rica and Cameroon have already started to manage tropical forests in a sustainable way by being very selective about which trees are cut, developing new forest products, and trying to reduce damage to seedlings, saplings and poles as much as possible. Reduced-impact logging research including skyline cable systems and helicopter logging, both of which reduce soil impacts compared to ground skidding with crawler tractors, is under way at the Center for International Forestry Research (Drs. Dennis Dykstra and F.E. Pulz, personal communication). However, these methods are costly and difficult to apply in tropical countries. The Balloon-logging method might be an alternative if prototypes that can lift heavy logs are designed (Conway 1982; Olsen et al. 1984). In certain cases, hand-logging by local villagers and animal-logging (e.g., tamed elephant) might be appropriate.

Increasing reforestation activities with fast-growing trees and promotion of more efficient cooking stoves and improved methods of charcoal production will reduce pressures on natural forests (Khasa et al. 1994, 1995). The monoculture of fast-growing exotics is often criticized, because of apparent lack of adaptation of the exotics used and their susceptibility to pests. At the same time, the long-term environmental impacts of the introduction of exotics are often unknown. Therefore, promotion of native species in reforestation programs should be encouraged because they are often better adapted to local growing conditions and contribute to the maintenance of natural biodiversity while providing food and habitats for local wildlife. On the other hand, the developments of hydraulic, solar, methane and biogas energies should be encouraged if they are socio-economically, culturally and environmentally acceptable and with minimum impacts on biodiversity.

Sustaining forest biodiversity requires well-trained personnel who know about the appropriate technologies and who also understand the institutional, socio-economic, and cultural aspects. While holding at least 50% of the world's species, there are only about 138 universities and 220 technical schools in tropical nations providing forestry education and training (US Congress OTA 1992). There are now two international centers: International Center for Research in Agroforestry in Nairobi (ICRAF) and Center for International Forestry Research in Bogor (CIFOR). Both are committed to developing strategic research in sustainable agriculture and forestry while applied research could be conducted in national institutions. Both fundamental and applied research should be underway to address the multiple problems related to a proper sustainable management of tropical forests. Potentially effective and profitable techniques should be disseminated among scientists, between scientists and technology users, and among technology users. Programs of public education and awareness of biodiversity, its importance, and the measures necessary for conserving it should be encouraged. Article 1 of the Convention on Biodiversity refers to the sustainable use and the fair and equitable sharing of the benefits arising out of the utilization of genetic resources. In this matter, the Merck & Co. Ltd. USA and the National Biodiversity Institute (INBIO) of Costa Rica agreement serves as an example.

CONCLUSION

Tropical forest biodiversity is work sustaining. Development of easily-applied biodiversity measures are essential to the prudent management for biodiversity. A number of strategies for each country are required to achieve sustainable living. These include preparation of an inventory of forest resources, management of natural forests, education and involvement of local communities, population control, integration of forest policies within wider policies, and prevention of pollution. The underlying causes of loss of forest biodiversity are institutional, social, and economic. Consequently, the reforms needed to support sustainable resource development should ultimately come from the governments and people of the tropical nations. The role of NGOs which are closer to the people, less bureaucratic and more innovative will have an important impact on biodiversity management if adequate financial support through international agencies is provided. Finally, the emergence of regional cooperative efforts such as those in the Association of Southeast Asian Nations (ASEAN) and the Southern African Development Community (SADC), the establishment of internationally networked programmes, such as the Canadian Model Forests and the Commonwealth Forestry Initiatives, and other new incentives, in socio-economic, cultural, political and ecological conditions of the area concerned, are likely to render the whole process of sustainable management of biodiversity more effective (Brand et al. 1993; Khasa and Dancik 1997).

REFERENCES

Aronoff, S. 1989. Geographic Information Systems: A Management Perspective. WDL Publications, Ottawa, Canada.

Bouman O. T. and D. G. Brand. 1997. Editorial Summary. In: Sustainable Forests: Global Challenges and Local Solutions. Proceedings of an International Conference. O.T. Bouman and D.G. Brand (eds.). Saskatoon, Saskatchewan, Canada, May 29-June 1, 1995.

Brand, D. G., R. W. Roberts and R. Kemp. 1993. International initiatives to achieve sustainable management of forests: Canada's model forests, the Commonwealth forestry initiative, and the development assistance community. Common For Rev. 72:297-302.

Burton, P. J., A. C. Balisky, L. P. Coward, S. G. Cumming and D. D. Kneeshaw. 1992. The value of managing for biodiversity. For. Chron. 68:225-237.

Caughley, G. 1994 Directions in conservation biology. J. Anim. Ecol. 63:215-244.

Conway, S. 1982. Logging practices: principles of timber harvesting systems. Miller Freeman Publications, Inc., California.

Edwards, C. A., R. Lal, P. Maddenn, R. H. Miller and G. House (eds.). 1990. Sustainable agricultural systems. Soil and Water Conservation Society, St. Lucie Press, Delray Beach, Fla.

Gillison, A. N. and G. Carpenter. 1994. A generic Plant Functional Attribute Set and Grammar for Vegetation Description and Analysis. CIFOR Working Paper no. 3.

Goulding, M. 1993. Flooded forests of the Amazon. Scientific American (March) 114-120.

Hendrix, P. F., D. A. Crossley, J. M. Blair and D. C. Coleman. 1990. Soil biota as components of sustainable agroecosystems. In: Sustainable agricultural systems. C. A. Edwards, R. Lal, P. Madden, R. H. Miller and G. House (eds.). pp. 637-654. Soil and Water Conservation Society, St. Lucie Press, Delray Beach, Fla.

Holdgate, M. 1993. Sustainability in the forest. Common. For. Rev. 72:217-225.

Katz, L. S., Orrick, S. and Honig, R. 1993. Environmental profiles: a global guide to projects and people. Garland Publishing, Inc. New York & London.

Khasa, P. D., G. Vallée and J. Bousquet. 1994. Biological considerations in the utilization of *Racosperma auriculiforme* and *R. mangium* in tropical countries with emphasis on Zaire. J. Trop. For. Sci. 6:422-443.

Khasa, P. D. and B. P. Dancik. 1997. Managing for Biodiversity in Tropical Forests. J. Sust. For. 4(1/2): 1-31.

Khasa, P. D., G. Vallée, J. Bélanger and J. Bousquet. 1995. Utilization and management of forest resources in Zaire. For. Chron. 71:479-488.

Kimmins, J. P. 1991. Biodiversity: an environmental imperative. In: Proceedings of the Canadian Public and Paper Association Meeting. pp. 225-231. Montreal, PQ.

Klier, K., M. J. Leoschke and J. F. Wendel. 1991. Hybridization and introgression in white and yellow ladyslipper orchids (*Cypripedium candidum* and *C. pubescens*). J. Heredity 82:305-318.

Kubitzki, K. and A. Ziburski. 1994. Seed dispersal in flood plain forests of Amazonia. Biotropica 26:30-43.

Ledig, F. T. 1988. The conservation of diversity in forest trees: why and how should genes be conserved? BioScience 38:471-479.

Li, P., J. Mackay and J. Bousquet. 1992. Genetic diversity in Canadian hardwoods: implications for conservation. For. Chron. 68:709-719.

Mandava, N. B. 1985 (ed.). CRC Handbook of Natural Pesticides. CRC Series in Naturally Occurring Pesticides. CRC Press, Inc., Boca Raton, Fla.

Marshall, D. R. 1990. Crop genetic resources: current and emerging issues. In: Plant Population Genetics, Breeding, and Genetic Resources. A. H. D. Brown, M. T. Clegg, A. L. Kahler and B. S. Weir (eds.). pp. 367-388. Sinauer, Sunderland, Mass.

Myers, N. 1989. Tropical forests and their species going, going? In: Biodiversity. E. O. Wilson and F. M. Peter (eds.). pp. 28-35. National Academy Press, Washington, DC.

Namkoong, G. 1991. Maintaining genetic diversity in breeding for resistance in forest trees. Annu. Rev. Phytopathol. 29:325-342.

Ojeda, F., J. Arroyo and T. Maranon. 1995. Biodiversity components and conservation of Mediterranean heathlands in southern Spain. Biol. Conserv. 72:61-72.

Olsen, E. D., B. L. Tuor and M. R. Pyles. 1984. Balloon logging with the pendulum-swing system: factors affecting lift. Forest Research Laboratory, Oregon State University, Corvallis. Research Bulletin 47.

Panayotou, T. and Ashton, P. S. 1992. Not by timber alone: economics and ecology for sustaining tropical forests. Island Press, Washington, DC, USA.

Parresol, B. R. and J. McCollum. 1997. Characterizing and comparing landscape diversity using GIS and a contagion index. J. Sust. For. 5(1/2): 249-261.

Pendje, G. and M. Baya. 1992. La réserve de biosphère de Luki (Mayobe, Zaire): patrimoine floristique et faunique en péril. UNESCO, Paris.

Pimentel, D., U. Stachow, D. A. Takacs, H. W. Brubaker, A. R. Dumas, J. J. Meaney, J. A. S. O'Neil, D. E. Onsi and D. B. Corzilius. 1992. Conserving biological diversity in agriculture/forestry systems. BioScience 42:354-362.

Ragland, J. and R. Lal. (eds.). 1993. Technologies for sustainable agriculture in the tropics ASA, CSSA AND SSSA, Madison, WI, USA.

Reddy, M. S., L. M. Funk, D. C. Covert, D. N. He and E. A. Pedersen. 1997. Microbial inoculants for sustainable forests. J. Sust. For. 5(1/2): 293-306.

Reid, W. V., J. A. McNeely, D. B. Tunstall, D. A. Bryant and M. Winograd. 1993. Biodiversity Indicators for Policy-Makers. World Resources Institute, New York.

Silbaugh, J. M. and D. R. Betters. 1997. Biodiversity values and measures applied to forest management. J. Sust. For. 5(1/2): 235-248.

Singh, K. D. 1993. The 1990 tropical forest resources assessment. Unasylva 44:10-19.

Steinlin, H. 1994. The decline of tropical forests. Quarterly Journal of International Agriculture 33:128-137.

Stoll, G. 1988. Protection Naturelle des Végétaux en Zones tropicales. Editions Josef Margraf, Weikersheim, Germany.

Trochain, J. L. 1980. Écologie végétale de la zone intertropicale non désertique. Université Paul Sabatier, Toulouse.

UNCED (The United Nations Conference on Environment and Development). 1992. Rapport de la conférence des nations unies sur l' environment et le développement. 3-14 June 1992, Rio de Janeiro. Vol. I, II, III, IV & V.

US Congress OTA (US Congress Office of Technology Assessment). 1992. Technologies to Sustain Tropical Forest Resources and Biological Diversity, OTA-F-515. Washington, DC. US Government Printing Office.

Veitch, N., N. R. Webb and B. K. Wyatt. 1995. The application of geographical information systems and remotely sensed data to the conservation of heathland fragments. Biol. Conserv. 72:91-97.

WCMC (The World Conservation Monitoring Centre). 1992. Global Biodiversity: Status of the Earth's Living Resources. Chapman and Hall, London.

Williams, P. H., K. J. Gaston and C. J. Humphries. 1994. Do conservationists and molecular biologists value differences between organisms in the same way? Biodiversity Letters 2:67-78.

Wilson, E. O. 1989. The current state of biological diversity. In: Biodiversity. E. O. Wilson and F. M. Peter (eds.). pp. 3-18. National Academy Press, Washington, DC.

Witting, L. and V. Leschcke. 1995. The optimization of biodiversity conservation. Biol. Conserv. 71:205-207.

Zobel, B. and J. Talbert. 1984. Applied Forest Tree Improvement. John Wiley & Sons, New York.

Biodiversity Values and Measures Applied to Forest Management

John M. Silbaugh
David R. Betters

ABSTRACT. With the new emphasis in land management agencies (particularly the USDA Forest Service) on ecosystem management, new attention is being paid to measurements of biological diversity (biodiversity). Forest managers faced with incorporating the maintenance of biodiversity into their analyses, plans, and decisions need reliable, quantitative biodiversity measures. Deciding which measures to use requires land managers to understand why biodiversity is valued, which components of diversity are of most value, and how to measure these components feasibly. In this paper, the values associated with biodiversity and the principal indices available to forest managers are reviewed. Attention is drawn to the differing emphases of biodiversity indices–species richness, heterogeneity, and evenness– and to how these differences address different values. Potential applications for these indices in forest management plans–including ways to use habitat models to approximate biodiversity–are also discussed. *[Article copies available for a fee from The Haworth Document Delivery Service: 1-800-342-9678. E-mail address: getinfo@haworth.com]*

INTRODUCTION

With the new emphasis in land management agencies (particularly the USDA-Forest Service) on ecosystem management, new atten-

John M. Silbaugh is Graduate Research Assistant and David R. Betters is Professor in the Department of Forest Sciences, Colorado State University, Ft. Collins, CO, USA.

[Haworth co-indexing entry note]: "Biodiversity Values and Measures Applied to Forest Management." Silbaugh, John M., and David R. Betters. Co-published simultaneously in *Journal of Sustainable Forestry* (Food Products Press, an imprint of The Haworth Press, Inc.) Vol. 5, No. 1/2, 1997, pp. 235-248; and: *Sustainable Forests: Global Challenges and Local Solutions* (ed: O. Thomas Bouman, and David G. Brand) Food Products Press, an imprint of The Haworth Press, Inc., 1997, pp. 235-248. Single or multiple copies of this article are available for a fee from The Haworth Document Delivery Service [1-800-342-9678, 9:00 a.m. - 5:00 p.m. (EST). E-mail address: getinfo@haworth.com].

235

tion is being paid to measurements of biological diversity (biodiversity). The need for reliable, quantitative measures of a rather fuzzy concept should be readily apparent. Scores of biodiversity indices are used by ecologists to monitor the effects of pollution or to designate conservation areas, usually through analysis of plots too small to be practical for forest management. Each of these measures, from species richness to heterogeneity to evenness, emphasizes slightly different components of biodiversity over others. Besides introducing confusion when choosing which index to use, the often contradictory results of these indices provide ample opportunity for their misuse.

Measuring (and managing) forest biodiversity requires an understanding of several important points or themes. These include (1) the need to express biodiversity quantitatively; (2) the importance of isolating the various values attached to biodiversity; (3) the relationships among the three distinct (and at times conflicting) components of species diversity—richness, heterogeneity, and evenness; and (4) the recognition of constraints of management feasibility (particularly concerning data availability) that necessitate in some instances accepting models and other approximations of biodiversity.

These themes are clearly interrelated in the quest for better understanding of what biodiversity is, why it is important, and how it can be monitored and managed. Public natural resource management planning in the United States requires many steps, with each step subject to appeal by industrial and environmental interests. To avoid expensive legal challenges and, more importantly, to provide the best management, the goals of planning and management must be defined as precisely as possible in order that multiple alternatives may be evaluated as objectively as possible. As biodiversity values are being integrated into the plans for U.S. National Forests and other federal lands, the need to quantify these values becomes increasingly acute (Loomis 1993). Using quantitative indices allows biodiversity objectives to be set out clearly, and alternative plans to be evaluated definitively.

Obstacles to quantification are numerous. Some spring from a fairly simple problem: biodiversity is often defined too broadly. Early definitions of biodiversity stressed the variety of organisms—in other words, species diversity. More recent descriptions are much more extensive, incorporating such elements as genetic variation within species, diversity of ecosystem "types," and structural and functional components of diversity (Franklin et al. 1981; Franklin 1988; Noss 1990; Wilson 1992). As these definitions expand, they become less useful for managers–virtually everything in the forest is seen as indicative of "biodiversity" (McKenney et al. 1994).

Given this impulse, it is easy to see why quantifying biodiversity is a daunting task. The added components are often confusing, frequently unmeasurable, and sometimes conflicting; successful measurement of biodiversity requires us to select the few that are of primary concern. The first step in that process is isolating the values of biodiversity that society (and science) see as most important.

VALUES AND MEASURES OF BIODIVERSITY

Humans attach to biodiversity many different values, separated here into three categories: biodiversity as a means to an end, as a measure of environmental quality, and as an end in itself. In the first category, biodiversity is valued for its contribution to the sustainable production of goods and services. The discovery of new medicines is an oft-cited example: Farnsworth (1988) lists more than 100 pure chemical substances extracted from plants and used worldwide for medicine, and Eisner (1990) reports that nearly one-quarter of all U.S. prescriptions are for formulas based on plant or microbe products or derivatives.

The diversity of forest species supplies many amenities and commodities–not only medicines, but foods, wood products, natural fibers, recreation opportunities, aesthetic benefits, ecological filtering, and more (Tobin 1990; Ehrlich and Wilson 1991; McMinn 1991; Burton et al. 1992; West 1993). If the goal of biodiversity conservation is to keep intact the diverse pieces of the supply base–in order to keep these forest goods and services flowing–then clearing biodiversity is valued instrumentally as a means to this end.

Predicting the chemical (and economic) potential of a particular species is, of course, difficult. It is unlikely that *every* plant, animal, or microbe will yield new products beneficial to humans. The implicit assumption in arguments for saving biodiversity based on commodity production, then, is that our lack of knowledge–about the number of species, their potential uses, and the future economic value of those uses–justifies preserving as many different species as we can.

In the second category of values, biodiversity serves as an early warning system by indicating environmental damage. Biological indicators are organisms so closely tied to certain environmental conditions that their presence indicates the existence of those conditions (Patton 1987). Pollution studies have used a wide range of organisms, from lichens to aquatic fly larvae to honeybees, field mice, and shrews, for biological monitoring (Root 1990). By allowing us to avoid polluting the environment to such a degree that it becomes uninhabitable for all life, these indicators are like the proverbial canary in a coal mine (Burton et al. 1992). Moreover, there is a sense that biodiversity, in and of itself, may indicate a desirable environment. In the words of one Forest Service researcher, high biodiversity reflects a diversity of ecological options, "the result of abundant opportunities for different organisms to acquire and use resources" (Robert J. Laacke pers. comm.[1]). As long as biodiversity is maintained, we have evidence that these opportunities are still available in our environment.

While conservation arguments based on its contributions to a better life for humankind are compelling, biodiversity is increasingly valued not for what it *contributes* but for what it *is*–this is the third category of biodiversity values. Whether the impetus is moral, religious, philosophical, or historical, many view biodiversity as an end-in-itself, quite apart from the instrumental uses it may have in making human lives better. Philosophers and environmental ethicists ascribe "intrinsic value" to entities that are valued as ends in themselves, rather than as the means to some other purpose (Marietta 1991; Rolston 1994). Unlike its value as a means to an end, the intrinsic value of a species is presumed to be equal to that of any other species. Piñon pine (*Pinus edulis*), though less commercially-valuable than lodgepole pine (*P. contorta*), has no less intrinsic value. The same is true for species' value as part of our natural history. The leading proponent of this historical value is E. O. Wilson, who suggests that each new species of higher organism discovered is "richer in information than a Caravaggio painting, Bach fugue, or any other great work of art" (Wilson 1985, p. 701).

These three categories of value not only provide a variety of justifications for conserving biodiversity but they also suggest, because of their distinct emphases, that different components of species diversity may be more essential depending on which values are paramount. These components include species richness, heterogeneity, and evenness.

Measures of Species Diversity

Species diversity indices have been useful in allowing ecologists to spot environmental problems and identify important areas for nature reserves (Magurran 1988). As Hill (1973), Peet (1974), and others have shown, the various species diversity measures differ primarily in their sensitivity to changes in the proportional abundance of rare vs. common species. Species richness measures, which are simple counts of the number of species present in an area, are widely-used because of their ease of computation. However, their ultra-sensitivity to the number of rare species present ensures that as the size of a sampling area increases (or, similarly, as the sampling effort on a given area increases), species richness increases accordingly (Magurran 1988; Wilson 1992).

Heterogeneity measures, the most well-known of which are Simpson's Index (Simpson 1949) and Shannon's Index (Shannon and Weaver 1964), are less sensitive to the number of rare species present and thus to the extent of the sampling area. In these measures, diversity is the sum of the weighted proportional abundances of all species found in the sample; diversity increases as rare species become less rare or as the most dominant species become less dominant (Hurlbert 1971).

The third type of species diversity measures—evenness measures—concentrate on the distribution of species' proportional abundances, independent of the total number of species present. Models of species abundance such as the geometric series, log series, and log normal distributions are measures of evenness, as are indices, such as Pielou's J statistic, which compare real diversity to a hypothetical maximum (May 1975; Pielou 1975; Magurran 1988). Evenness measures are the least sensitive of the three types to sampling intensity, driven more by the proportional abundance of the most dominant species than by the number of species found.

While these three measures of species diversity have been used widely in ecological plot studies, with the exception of species richness, they have rarely been applied at larger geographic scales. One obvious obstacle to using these measures at forest scales is the amount of data collection necessary to derive, for example, a heterogeneity index for birds on a 200,000-acre forest. Instead of measuring species diversity directly, then, forest managers are likely to try to gauge forest biodiversity by the performance of certain high-profile vertebrate species (often threatened or endangered), or, increasingly, by the patterns of landscape patches distributed across regional scales (Noss 1983). In either case, we find these efforts inadequate to address biodiversity. In the next section, we offer alternatives for forests that are based on the fine-scale species diversity measures described above.

RECOMMENDATIONS FOR FUTURE STUDY AND SPECIFIC APPLICATIONS

The reluctance of some researchers to use species diversity measures is due to the sometimes contradictory results they can yield. Efforts to construct intrinsic diversity profiles (Grassle et al. 1979; Lewis et al. 1988; Gove et al. 1992) for comparing communities arose from this phenomenon. However, as each measure emphasizes a different component of species diversity, forest managers can and ought to choose the specific measure(s) best corresponding to the forest's values-based objectives.

Variety Equals Sustainability

For example, in order to ensure the continued production of many resources (mandated by the Multiple Use-Sustained Yield Act of 1960), a U.S. National Forest must take steps to protect a varied production base of plants, animals, fungi, microbes, etc. This variety is essentially species richness. If maintaining variety is set out as a management objective, then areas of greatest species richness would be appropriate areas for protection. Likewise, protecting areas rich in threatened and endangered species also focuses conservation strategy on the maintenance of a variety of species, thus protecting the forest's potential sustainability.

Because a sustainable supply of forest products may ultimately depend on maintaining genetic variety, many techniques have been proposed for enhancing the diversity of genes within a population (e.g., Probst and Crow 1991; Kemp 1992). A similar rationale is used to suggest that conservation of individual species be prioritized so as to maximize genetic diversity. For example, this approach has been proposed in connection with conservation in the crane family (Solow et al. 1993; Weitzman 1993). The chief difficulty with this strategy for forest managers (apart from ethical questions about "playing God") is the extraordinary cost and effort associated with individual species conservation, especially when critical habitat for the species lies outside the forest's boundaries.

Whether the focus is on the species level or on the genetic level, protecting sustainability means protecting rarity. From the perspective of biodiversity-for-sustainability, the value of any species is ultimately derived from the actual and potential products and services it provides. Thus, assuming conservation biology can keep a rare species viable long enough for the discovery of a medicine, food, or unique recreational or aesthetic benefit associated with that species, the marketplace for these benefits will ensure the species' further protection. This fundamental trust in the free market system is one of the underlying (and usually unstated) assumptions behind arguments for protecting biodiversity for the sake of sustainability. Other perspectives regarding biodiversity exhibit much less trust in economics.

Measuring Environmental Quality

If biodiversity is valued less as a means to sustainability than as an indicator of environmental quality, then species richness measures are probably not appropriate. At the site level, changes in the relative abundances of species, not in the overall number of species, are the first indicators of environmental degradation (Magurran 1988). Evenness measures offer greater sensitivity to environmental contamination than either richness or heterogeneity measures. (In fact, tipping off such contamination may be the *only* real use for evenness measures.) Other proposed site-level environmental quality measures, such as monitoring nutrient cycling rates (Noss 1990), seem to pertain less to biodiversity than to the "normal" functioning of ecosystems. While it is likely that these rates play a role in the maintenance of species diversity, documenting the nature of the role is difficult. In any event, using site-level process rates as a measure of biodiversity makes little sense for forest management, again because of the huge data collection required.

In fact, most attempts to tie biodiversity to environmental quality at the forest level are problematic. For example, while indicator species have a well-documented role in revealing air and water pollution (Root 1990; Tobin 1990), their presence alone shouldn't be equated with biodiversity. Alternatively, recent research has focused on landscape disturbance frequencies as a measure of the naturalness of managed forests—how closely they parallel the corresponding historic spatial and temporal disturbance patterns (Turner et al. 1993; Reice 1994). Here again, it remains to be shown that these natural landscape patterns generate either higher species diversity or a higher-quality environment than less natural patterns.

Biodiversity as an End: Heterogeneity Measures and Habitat Models

Valuing biodiversity because of what it contributes to sustainability or environmental quality is certainly legitimate, but the links between means and ends are susceptible to fluctuating economic climates, cost-benefit analyses, and arbitrary and changeable standards of what constitutes environmental quality. Valuing biodiversity for its own sake provides a stronger, more lasting justification for the work of preserving it. Consequently, deciding which techniques best measure this value is particularly important and should be a top priority for forest management.

Maintaining biodiversity for its own sake is best accomplished via species heterogeneity measures within communities/ecosystems. More than either richness or evenness alone, heterogeneity truly reflects the effective diversity of a collection by incorporating *both* of these elements. This diversity has real effects in the forest, increasing the probability of interspecific encounters (Hurlbert 1971) and enriching the diversity of life that human observers will encounter when exploring the forest (Wilson 1992). Further, heterogeneity measures treat all species equally and thus incorporate fully the intrinsic value of each species and of biodiversity.

Within a forest, the best way to measure species heterogeneity is to use habitat rather than population data. This approach is both more feasible (population data aren't even available for many species) and more accurate. Population numbers will vary seasonally and annually as a result of stochastic variation–not necessarily because biodiversity is increasing or decreasing. Studies have even suggested that population numbers can mislead, particularly in territorial vertebrates, where better habitat is often less-populated than poorer sites (Van Horne 1983).

Habitat modelling serves a dual purpose, helping managers to better understand the biology of animal species while also providing a diversity measure. A drawback of currently available habitat models, though, is that only vertebrate animals are considered: plants, insects, and fungi are excluded from analysis. The emphasis on vertebrates in habitat modelling stems from their mobility; the farther an animal roams, the more necessary it becomes to identify areas (habitat) where the animal *could* be rather than where it *is*. Besides, plants and sessile invertebrates are more likely to be the focus of site-level environmental monitoring efforts (discussed above).

There are at least two ways in which habitat models can be used to yield measures of forest biodiversity. One is to translate acres of habitat into supportable numbers of individual animals, species-by-species. This is the approach of the USDA Forest Service's HABCAP (Habitat Capability) models, which link habitat to potential carrying capacity via per-acre multipliers that vary with the quality (optimal, suitable, etc.) of the habitat. Coming up with the multipliers requires incorporating the relative differences between species–body size, home range and territory size, social structures (herds, packs, breeding pairs, individuals)–into the models.

The second way to use habitat modelling to measure biodiversity dispenses entirely with estimates of population numbers and simply uses habitat acres (again, scaled to home range size) as a stand-in for population numbers. This method is preferred at one U.S. National Forest, partly because of the spurious connection between habitat quality and population numbers observed by Van Horne (1983) and partly because managing habitat is simply more feasible than managing animal populations (Robert J. Laacke pers. comm.[2]).

In both cases, attention to the spatial arrangement of habitat is crucial, especially for species with large home ranges and relatively low tolerances for traversing unsuitable habitat. For example, elk (*Cervus elaphus*) require large blocks of contiguous habitat; small, isolated patches of habitat are useless. For species with smaller ranges, like voles or salamanders, connectivity is certainly less important. Geographic Information Systems (GIS) technology has allowed for better spatial analysis of connectivity. Both the USDA Forest Service Pacific Northwest Region's HABSCAPES (Habitat Analysis of Landscapes) model and the State of California's WHR (Wildlife Habitat Relationships) system use grid-based nearest-neighbour algorithms to determine whether a single pixel of optimal habitat is an isolated fragment or part of a larger clump of similarly good habitat (Mellen et al. 1994; Robert J. Laacke pers. comm.[3]). As knowledge increases about species' ability to disperse from patch to patch through less-than-optimal habitat, these algorithms can be refined to accommodate this new information. A spatially-explicit dispersal model developed to study northern spotted owls (*Strix occidentalis caurina*) in Washington and Oregon (called, appropriately, OWL) represents just such a refinement.

The final key to making habitat modelling work for biodiversity analysis is to be thoughtfully selective about which species to analyze. Technically, it is possible to evaluate habitat for hundreds of vertebrate species on every 30-meter-square pixel of land in a forest, to assess each species' home range and territory limits, to apply carrying capacity estimates based on this habitat analysis, and to arrive at some idea of the potential populations for each species in the forest. This would yield a static picture of vertebrate species heterogeneity, along with a hefty bill for computer time. Whereas the primary interest in measuring biodiversity is to *manage* it, however, a static picture is insufficient. The forest planner then must evaluate the changes in those potential populations based on different management alternatives; given half a dozen such alternatives and a planning horizon of 10 to 50 years, trying to look at *all* species at once is probably impractical.

Selecting species that effectively "stand in" for others (often called "management indicator species") can be helpful (Patton 1987). However, the manager must be certain that the indicators really do represent the habitat requirements of their proxy species. In the

HABSCAPES model, for example, species are grouped into "guilds" based on their home range size, seral stage preference, and dependence on contrast (edge) or homogeneous (interior forest) habitat. The habitat analysis is then run for the entire guild sharing the same preferences (Mellen et al. 1994). Another way of selecting species for analysis would be to use qualitative distinctions based on their functions in ecosystems–identifying "key stone" species, for example–again being careful to ensure that solid research supports these qualitative designations (Cousins 1991; Noss 1991).

Habitat models represent the crucial link between fine-scale heterogeneity and forest-scale diversity measurement. As it becomes increasingly clear that society does value biodiversity as an end in itself, expanding the use of these models and refining their applications takes on a greater urgency. The limits of our ability to collect enough data to understand the current status and trend of every species determine what is feasible to manage. Using habitat models–approximations of reality–should help managers make the best decisions given the available data.

CONCLUSION

If biodiversity measures are to be useful to forest managers, they must be meaningful (both biologically and socially), measurable (quantitative and obtainable from available data), and manageable (subject to change based on human decisions). Separating the many components of biodiversity, keeping in sight the reasons we value biodiversity, and recognizing what is feasible and cost-effective will go a long way toward helping forest managers incorporate biodiversity into their management plans.

NOTES

1. Robert J. Laacke, USDA Forest Service, Pacific Southwest Forest & Range Experiment Station, Silviculture Laboratory, Redding, CA; letter dated November 29, 1994.
2. Ibid.
3. Telephone conversation. April 28, 1995.

REFERENCES

Burton, P. J., A. C. Balisky, L. P. Coward, S. G. Cumming, and D. D. Kneeshaw, 1992. "The Value of Managing for Biodiversity," Forestry Chronicle 68(2), 225-236.

Cousins, S. H., 1991. "Species Diversity Measurement: Choosing the Right Index," Trends in Ecology and Evolution 6(6), 190-192.

Ehrlich, P. R. and E. O. Wilson, 1991. "Biodiversity Studies: Science and Policy," Science 253, 758-762.

Eisner, T., 1990. "Prospecting for Nature's Chemical Riches," Issues in Science and Technology 6(2), 31-34.

Farnsworth, N. R., 1988. "Screening Plants for New Medicines," pp. 83-97 in Wilson, E. O. and F. M. Peters, eds., Biodiversity, National Academy Press, Washington, DC.

Franklin, J. F., K. Cromack, Jr., W. Denison, A. McKee, C. Maser, J. Sedell, F. Swanson, and G. Juday, 1981. "Ecological Characteristics of Old-Growth Douglas-Fir Forests," USDA Forest Service, General Technical Report PNW-118. Pacific Northwest Forest & Range Experiment Station, Portland, OR.

Franklin, J. F., 1988. "Structural and Functional Diversity in Temperate Forests," pp. 166-175 in Wilson, E. O. and F. M. Peters, eds., Biodiversity, National Academy Press, Washington, DC.

Gove, J. H., C. W. Martin, G. P. Patil, D. S. Solomon, and J. W. Hornbeck, 1992. "Plant Species Diversity on Even-aged Harvests at the Hubbard Brook Experimental Forest: 10-year Results," Canadian Journal of Forest Research 22, 1800-1806.

Grassle, J. F., G. P. Patil, W. Smith, and C. Taillie, eds. 1979. Ecological Diversity in Theory and Practice. International Cooperative Publishing House, Fairland, MD.

Hill, M. O., 1973. "Diversity and Evenness: A Unifying Notation and Its Consequences," Ecology 54, 427-432.

Hurlbert, S. H., 1971. "The Nonconcept of Species Diversity: A Critique and Alternative Parameters," Ecology 52, 577-586.

Kemp, R. H., 1992. "The Conservation of Genetic Resources in Managed Tropical Forests," Unasylva 169 43, 34-40.

Lewis, C. E., B. F. Swindel, and G. W. Tanner, 1988. "Species Diversity and Diversity Profiles: Concept, Measurement, and Application to Timber and Range Management," Journal of Range Management 41(6), 466-469.

Loomis, J. B., 1993. Integrated Public Lands Management: Principles and Applications to National Forests, Parks, Wildlife Refuges, and BLM Lands. Columbia University Press, New York, NY.

Magurran, A. E., 1988. Ecological Diversity and Its Measurement. Princeton University Press, Princeton, NJ.

Marietta, D. E., Jr., 1991. "Thoughts on the Taxonomy and Semantics of Value Terms," Journal of Value Inquiry 25, 43-53.

May, R. M., 1975. "Patterns of Species Abundance and Diversity," pp. 81-120 in

Cody, M. L. and J. M. Diamond, eds., Ecology and Evolution of Communities. Belknap Press of Harvard University Press, Cambridge, MA.

McKenney, D. W., R. A. Sims, M. E. Soulé, B. G. Mackey, and K. L. Campbell, eds., 1994. "Toward a Set of Biodiversity Indicators for Canadian Forests: Proceedings of a Forest Biodiversity Indicators Workshop," Natural Resources Canada, Canadian Forest Service, Sault Ste. Marie, ON.

McMinn, J. W., 1991. "Biological Diversity Research: An Analysis," USDA Forest Service, General Technical Report SE-71. Southeastern Forest Experiment Station, Asheville, NC.

Mellen, K., M. Huff, and R. Hagestedt, 1994. "Interpreting Landscape Patterns: A Vertebrate Habitat Relationships Approach," Introduction to HABSCAPES (Habitat Analysis of Landscapes) computer model, available from USDA Forest Service, Mt. Hood National Forest, Gresham, OR.

Noss, R. F., 1983. "A Regional Landscape Approach to Maintain Diversity," BioScience 33(11), 700-706.

Noss, R. F., 1990. "Indicators for Monitoring Biodiversity: A Hierarchical Approach," Conservation Biology 4(4), 355-364.

Noss, R. F., 1991. "From Endangered Species to Biodiversity," pp. 227-246 in Kohm, K. A., ed., Balancing on the Brink of Extinction: The Endangered Species Act and Lessons for the Future, Island Press, Washington, DC.

Patton, D. R., 1987. "Is the Use of 'Management Indicator Species' Feasible?" Western Journal of Applied Forestry 2(1), 33-34.

Peet, R. K., 1974. "The Measurement of Species Diversity," Annual Review of Ecological Systems 5, 285-307.

Pielou, E. C., 1975. Ecological Diversity. Wiley-Interscience, New York, NY.

Probst, J. R and T. R. Crow, 1991. "Integrating Biological Diversity and Resource Management," Journal of Forestry 89(2), 12-17.

Reice, S. R, 1994. "Nonequilibrium Determinants of Biological Community Structure," American Scientist 82, 424-435.

Rolston, H., III, 1994. Conserving Natural Value. Columbia University Press, New York, NY.

Root, M., 1990. "Biological Monitors of Pollution," BioScience 40, 83-86.

Shannon, C. E. and W. Weaver, 1964. The Mathematical Theory of Communication. University of Illinois Press, Urbana, IL.

Simpson, E. H., 1949. "Measurement of Diversity," Nature 163, 688.

Solow, A., S. Polasky, and J. Broadus, 1993. "On the Measurement of Biological Diversity," Woods Hole Oceanographic Institution mimeo in Journal of Environmental Economics & Management 24(1), 60-68.

Tobin, R. J., 1990. The Expendable Future: U.S. Politics and the Protection of Biological Diversity. Duke University Press, Durham, NC.

Turner, M. G., R. H. Gardner, R. V. O'Neill, and S. M. Pearson, 1993. "Multiscale Organization of Landscape Heterogeneity," pp. 81-87 in Jensen.

M. E. and P. S. Bourgeron. eds., Eastside Forest Ecosystem Health Assessment—Volume II: Ecosystem Management: Principles and Applications, USDA Forest Service, Pacific Northwest Research Station, Portland, OR.

Van Horne, B., 1983. "Density as a Misleading Indicator of Habitat Quality," Journal of Wildlife Management 47(4), 893-901.

Weitzman, M. L., 1993. "What to Preserve? An Application of Diversity Theory to Crane Conservation," Quarterly Journal of Economics CVIII, 157-183.

West, N. E., 1993. "Biodiversity of Rangelands," Journal of Range Management 46(1), 2-13.

Wilson, E. O., 1985. "The Biological Diversity Crisis," BioScience 35(11), 700-706.

Wilson, E. O., 1992. The Diversity of Life. Harvard University Press, Cambridge, MA.

COMPUTER MODELS CITED

HABCAP: Habitat Capability Model for the Rocky Mountain Region. Developed by Bill Aney, USDA Forest Service, Black Hills National Forest, Spearfish, SD. Documentation and User's Guide available from USDA Forest Service, Rocky Mountain Region, Lakewood, CO.

HABSCAPES: Wildlife Habitat Relationships Database and Programs for the Pacific Northwest. Developed by Kim Mellen and Rich Hagestedt, USDA Forest Service, Mt. Hood National Forest, Gresham, OR, and Mark Huff, USDA Forest Service, Pacific Northwest Research Station, Portland, OR. Documentation and User's Guide available from the developers.

OWL: Spatially-explicit Simulator for Northern Spotted Owl. Developed by Kevin McKelvey, USDA Forest Service, Pacific Southwest Forest & Range Experiment Station, Redwood Sciences Laboratory, Arcata, CA. Contact the developer for more information.

WHR: California Wildlife Habitat Relationships System. Developed by Reginald Barrett and Irene Timossi, California Interagency Wildlife Task Group. Documentation and User's Guide available from Barry Garrison, California Department of Fish and Game, Sacramento, CA.

Characterizing
and Comparing Landscape Diversity
Using GIS and a Contagion Index

Bernard R. Parresol
Joseph McCollum

ABSTRACT. The purpose of this study was to examine the pattern and changes in forest cover types over the last two decades on three landscape level physiographic provinces of the state of Alabama, USA: (i) The Great Appalachian Valley Province, (ii) The Blue Ridge Talladega Mountain Province, and (iii) The Piedmont Province. Studies of spatial patterns of landscapes are useful to quantify human impact, predict wildlife effects or describe various landscape features. A robust landscape index should quantify two distinct components of landscape diversity: composition and configuration. Composition refers to both the total number of "patch" types (i.e., forest cover types) and their relative proportions in the landscape, whereas configuration refers to the spatial pattern of patches in the landscape. The U.S. Forest Service conducts periodic surveys of forest resources nationwide from plots distributed on a 3 mile by 3 mile (4.8 km by 4.8 km) grid randomly established within each county using forest inventory and analysis survey data stratified by physiographic province, a relative contagion (RC) diversity value and its variance were calculated for each province for the survey years 1972, 1982, and 1990. One-way analysis of variance was used

Bernard R. Parresol and Joseph McCollum are affiliated with the USDA Forest Service, Southern Research Station, Asheville, NC, USA.

[Haworth co-indexing entry note]: "Characterizing and Comparing Landscape Diversity Using GIS and a Contagion Index." Parresol, Bernard R., and Joseph McCollum. Co-published simultaneously in *Journal of Sustainable Forestry* (Food Products Press, an imprint of The Haworth Press, Inc.) Vol. 5, No. 1/2, 1997, pp. 249-261; and: *Sustainable Forests: Global Challenges and Local Solutions* (ed: O. Thomas Bouman, and David G. Brand) Food Products Press, an imprint of The Haworth Press, Inc., 1997, pp. 249-261. Single or multiple copies of this article are available for a fee from The Haworth Document Delivery Service [1-800-342-9678, 9:00 a.m. - 5:00 p.m. (EST). E-mail address: getinfo@ haworth.com].

249

for hypothesis testing of RC values across time and between provinces. A view of each landscape at each point in time was generated with GIS software using Thiessen or proximal polygons of the forest cover types identified at each survey point on the landscape. *[Article copies available for a fee from The Haworth Document Delivery Service: 1-800-342-9678. E-mail address: getinfo@haworth.com]*

INTRODUCTION

Resource professionals have been expanding current management approaches to address landscape-level concerns and issues (Salwasser 1990). With the recognition of such problems as loss of biodiversity, climate change, and ecosystem degradation, a necessary evolution in management scale has come about. Today, resource managers are designing programs that consider the wellbeing of an ecosystem. The limitations of species-based management and narrowly focused project-by-project, permit-by-permit analysis have led to the development of geographically targeted approaches (Oliver 1992; Ticknor 1992). Based on watersheds or other ecological units rather than political boundaries, such approaches provide the tools and conceptual structures for ecosystem management and sustainable forestry (Council on Environmental Quality 1990).

Sustainable forestry involves the application of ecological/silvicultural principles, social/cultural values, and economic realities in long range planning as well as day to day decision making at regional, landscape, and local scales (Agee and Johnson 1988). To manage forest resources for sustainability, as with all types of management, requires the input of quantitative information. Assessing diversity at the landscape scale, valuable input for sustainable forest management, has recently gained much attention (Naveh and Lieberman 1994). Further developments are needed in the area of landscape measures of diversity, especially in terms of distributional properties so hypothesis tests can be conducted.

A robust landscape index should quantify two distinct components of landscape diversity: composition and configuration. Composition refers to both the total number of land use categories or "patch" types and their relative proportions in the landscape, whereas configuration refers to the spatial pattern of patches in the landscape (Li and Reynolds 1993). Contagion, as defined by

O'Neill et al. (1988), measures the extent to which landscape elements are aggregated or clumped. Higher values of contagion generally result from landscapes with a few large, contiguous patches, whereas lower values usually characterize landscapes with many small patches. Also, contagion values, in general, decrease as category proportions become more even. To date, only a handful of contagion indices have been proposed, and in no cases have distributional properties been examined.

Landscape diversity is inexorably linked to geographic information. Geographic Information Systems (GIS) are today an important tool for assessment, management, and monitoring. Layers of information can be combined in various ways to get a picture of the whole, or a visual perspective on change, or to highlight critical features, etc. For this paper geographic information, utilizing a GIS, is exploited to highlight regional features (physiographic provinces), landscape features (sample points and forest cover types), and for assessment.

MATERIALS AND METHODS

Physiographic Provinces and Forest Cover Type Data

To assess landscape habitat diversity one must necessarily specify the spatial boundaries of a landscape chosen for study. For geographically targeted units, the boundaries may coincide with those of an *island*; the word, in its ecological sense, connotes any environment entirely surrounded by another with strikingly different properties, the two being separated by a fairly abrupt boundary (Pielou 1975). Major patterns of landform-geologic material or "physiographic provinces" are well defined in the state of Alabama, USA (Hodgkins et al. 1979) (Figure 1). Three physiographic provinces were chosen for study: (i) The Great Appalachian Valley Province, (ii) The Blue Ridge-Talladega Mountain Province, and (iii) The Piedmont Province.

The Southern Forest Inventory and Analysis (SFIA) unit of the USDA Forest Service, Southern Research Station, conducts continuing inventories of forest resources in seven midsouth states. Data

FIGURE 1. State of Alabama, USA and its landscape physiographic provinces.

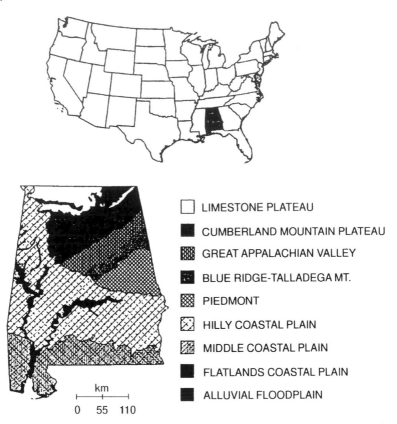

☐ LIMESTONE PLATEAU

■ CUMBERLAND MOUNTAIN PLATEAU

▨ GREAT APPALACHIAN VALLEY

▨ BLUE RIDGE-TALLADEGA MT.

▨ PIEDMONT

▨ HILLY COASTAL PLAIN

▨ MIDDLE COASTAL PLAIN

■ FLATLANDS COASTAL PLAIN

■ ALLUVIAL FLOODPLAIN

km
├──┼──┤
0   55   110

are collected from trees occurring on sample plots spaced across each state. Within each state plots are distributed on a 3- by 3-mile (4.8- by 4.8-km) grid randomly established within each county. Locations of survey plots, stratified by physiographic province, are shown in Figure 2. Because each physiographic province crosses many county lines, survey plots within a province do not form a systematic grid. Hence the area represented by a plot may be polygonal in shape instead of square or rectangular. From the tree data a forest cover type is identified for each plot. Forest cover types on

FIGURE 2. Distribution of SFIA survey plots within the three physiographic provinces.

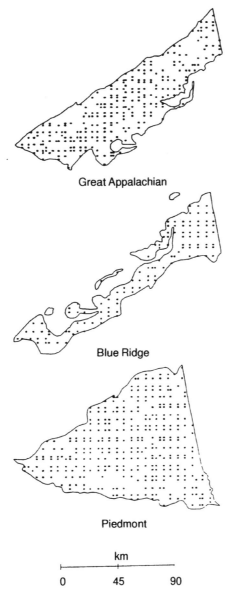

Great Appalachian

Blue Ridge

Piedmont

km

0   45   90

the study areas are listed in Table 1. Detailed descriptions of the SFIA data can be found in May (1990). In Alabama, the sixth forest survey was completed in 1990. The data from this and two previous surveys (1972 and 1982) were chosen for analysis.

Viewing Forest Cover Types on the Landscapes

Thiessen polygons can be used to apportion a point coverage into regions known as Thiessen of proximal polygons (Environmental Systems Research Institute 1992). Each region contains only one point and has the unique property that any location within a region is closer to the region's point than to the point of any other region. Each physiographic province point coverage (see Figure 2) was apportioned into Thiessen polygons and like polygons (polygons of the same forest cover type) were colored the same to create a landscape level view of forest cover types for each province at the three survey years (Figure 3).

TABLE 1. USDA Forest Service general forest cover types occurring on these physiographic provinces in the state of Alabama, USA

General forest cover types	Species
Nonforest	
Longleaf-slash pine	*Pinus palustris-Pinus elliottii*
Loblolly-shortleaf pine	*Pinus taeda-Pinus echinata*
Oak-pine	*Quercus* sp.-*Pinus* sp.
Oak-hickory	*Quercus* sp.-*Carya* sp.
Oak-gum-cypress	*Quercus* sp.-*Liquidambar styraciflua-Taxodium distichum*
Elm-ash-cottonwood	*Ulmus* sp.-*Fraxinus* sp.-*Populus* sp.
Non-typed	

FIGURE 3. Forest cover type changes through time in three physiographic provinces based on SFIA survey data.

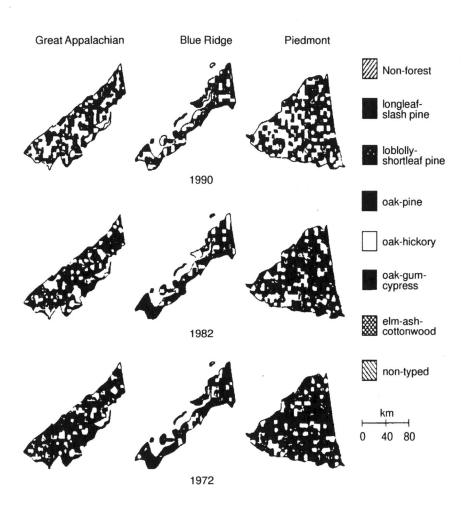

Great Appalachian   Blue Ridge   Piedmont

1990

1982

1972

Non-forest

longleaf-
slash pine

loblolly-
shortleaf pine

oak-pine

oak-hickory

oak-gum-
cypress

elm-ash-
cottonwood

non-typed

km

0   40   80

Contagion Index

The following contagion index, from Li and Reynolds (1993), quantifies both composition and configuration of the landscape:

$$RC = 1 + \sum_{i=1}^{n} \sum_{j=1}^{n} p_{ij} \ln(p_{ij})/2\ln(n)$$

and

$$p_{ij} = p_i p_{j/i}$$
$$p_{j/i} = N_{ij}/N_i$$

where p_{ij} is the probability that a polygon of land use i is found adjacent to a polygon of land use j and n is total number of land use categories. The components of p_{ij} are p_i, the probability that a randomly chosen polygon belongs to patch type i (estimated by the proportion of patch type i), and $p_{j/i}$, the conditional probability, where N_{ij} is the number of joins between polygons of patch types i and j, and N_i is the total number of joins between polygons of patch type i and all patch types (including patch i itself). RC stands for relative contagion. Values for RC lie between 0 and 1.

Contagion indices are normally computed on (square or rectangular) pixel based images. Joins are based on the four neighboring pixels. Since survey plots represent Thiessen polygonal areas, this application is a departure from the normal procedure. No effort was made to weight the count of joins based on join length or other suitable procedures. This is a refinement that should be looked into further.

Variance

While the above index is not new, the variance of RC has never been worked out. Using a similar approach as that used by Basharin (1959) for the variance of the classic entropy index, the variance of RC can be shown to be:

$$V(RC) = 1/2N\ln(n)\{\Sigma\Sigma p_{ij}\ln^2(p_j) - [\Sigma\Sigma p_{ij}\ln(p_{ij})]^2\} + O(N^{-2})$$

where N is number of plots and $O(N^{-2})$ is a remainder term of order of magnitude $1/N^2$. It can be shown that as N tends toward infinity the distribution of RC converges to the normal distribution. For those interested in the details of the variance derivation and distribution properties, please write to the first author.

Hypothesis Testing

Because an expression for the variance is available, and the distribution of the contagion index is normal (under large sample sizes, which we have), rigorous hypothesis testing can be accomplished through application of one-way analysis of variance (ANOVA). One-way ANOVAs can be constructed as follows: let t = number of groups, and T_i = number of observations for the ith group, then:

among groups, variance =

$$\sum_{i=1}^{t} (RC_i - \overline{RC})^2/(t - 1); \quad \overline{RC} = \sum_{i=1}^{t} T_i RC_i / \sum_{i=1}^{t} T_i$$

within groups, variance =

$$\sum_{i=1}^{t} T_i V(RC_i) / \sum_{i=1}^{t} T_i$$

F = among groups variance/within groups variance with $t - 1$,

$$\Sigma T_i - t \text{ degrees of freedom}$$

The formula just given for within groups variance is a pooled estimate of the common variance. If variances are not equal then weighted ANOVA should be used. In such a case, divide each RC value by the square root of its variance and compute among groups variance. Within groups variance is now 1, hence F reduces to among groups variance on the weighted RC values.

RESULTS AND DISCUSSION

It is generally believed that prior to the 1900s forested landscapes in the South were more homogeneous and contiguous than today. Exploitative logging, agriculture, forest type conversion, and other factors have altered, and continue to alter, the mosaic of forest cover types on the landscape. Today landscape flux (changes in composi-

TABLE 2. Contagion diversity values, variance and F-tests

Province	Year	RC	V(RC)
Great Appalachian Valley	1972	0.10495	.00049116
Great Appalachian Valley	1982	0.07092	.00036146
Great Appalachian Valley	1990	0.15306	.00069026
Blue Ridge-Talladega Mt.	1972	0.11693	.0015349
Blue Ridge-Talladega Mt.	1982	0.11700	.0014719
Blue Ridge-Talladega Mt.	1990	0.144337	.0010133
Piedmont	1972	0.15522	.00032129
Piedmont	1982	0.06879	.00025456
Piedmont	1990	0.21527	.00060121

H0: For Appalachian	$RC_{72} = RC_{82} = RC_{90}$;	F = 3.324,	Prob = 0.036
H0: For Blue Ridge	$RC_{72} = RC_{82} = RC_{90}$;	F = 0.188,	Prob = 0.829
H0: For Piedmont	$RC_{72} = RC_{82} = RC_{90}$;	F = 13.80,	Prob < 0.001
H0: For 1972	$RC_A = RC_B = RC_p$;	F = 9.287,	Prob < 0.001
H0: For 1982	$RC_A = RC_B = RC_p$;	F = 0.447,	Prob = 0.640
H0: For 1990	$RC_A = RC_B = RC_p$;	F = 5.212,	Prob = 0.006

tion and/or configuration) can occur on the time scale of a decade, as is readily seen in Figure 3 and Tables 2 and 3.

All *RC* values in Table 2 are fairly low, reflecting the fact that all three landscapes have many patches. Compositional change, in terms of changing proportions of forest cover types (see Table 3), is probably more responsible for the differences in *RC* values than configuration, except on the Piedmont. From the tests of hypotheses in Table 2 we can see that there was a significant increase in contagion in the Great Appalachian Valley in 1990 over the years 1982 and 1972. For the Blue Ridge-Talladega Mountain the increase in contagion for 1990 was not significant but in the Piedmont the contagion values were significantly different and followed a quadratic trend; up in 1972 then down in 1982 but with a notable rise in 1990. For the survey period 1972 the Piedmont had greater conta-

TABLE 3. Proportion of each forest cover type on each province by survey year

Forest cover type	Great Appalachian Valley			Blue Ridge-Talladega Mt.			Piedmont		
	72	82	90	72	82	90	72	82	90
				- - - - - - - - values in percent - - - - - - - -					
Non-forest	2.5	3.6	1.1	4.7	3.9	0.8	2.1	3.3	1.2
Longleaf-slash pine	33.2	28.5	8.3	39.1	37.5	10.5	27.2	23.9	7.4
Loblolly-shortleaf pine	13.7	14.6	24.8	11.7	10.9	22.6	19.7	17.6	24.0
Oak-pine	20.9	20.1	18.8	15.6	18.0	21.8	20.0	20.9	19.9
Oak-hickory	11.6	14.2	36.5	10.9	10.9	33.8	12.8	15.5	36.8
Oak-gum-cypress	18.1	19.0	10.5	18.0	18.8	10.5	17.9	18.8	10.4
Elm-ash-cottonwood							0.3		
Nontyped									0.3

259

gion over the other provinces but in 1982 all provinces had similar contagion. In 1990 the Piedmont again clearly had greater contagion than the Great Appalachian Valley or Blue Ridge-Talladega Mountain Provinces. From these tests we can generally conclude that (1) contagion is up in 1990, that is, landscape scale diversity has improved over the last decade and (2) the Piedmont Province has greater contagion (is less fragmented) than the other two provinces.

On all three landscapes there is a drastic loss (from 72 to 75 percent) of the longleaf-slash pine forest type between 1972 and 1990 (Table 3). The loblolly-shortleaf type appears to double in the Appalachian and Blue Ridge provinces while in the Piedmont it increases by about 20 percent. The oak-hickory type appears to triple across all provinces while the oak-gum-cypress type is dropping around 40 percent on the three provinces. From Figure 3, the oak-hickory and loblolly-shortleaf types are seen to occupy much of the former longleaf-slash and oak-gum-cypress types. In the southern U.S. longleaf pine had lost favor because of its "grass stage" where there is little aboveground growth for 3 to 6 years or more. Unfortunately, the other southern pines are more susceptible to fusiform rust disease, and species like the endangered red-cockaded woodpecker (*Picoides borealis*) prefer longleaf pine over the other southern pines. Prescriptions for hastening the passage through the grass stage have been developed and regeneration techniques have now been well documented (Barnett et al. 1990). Through the use of GIS, landscape views such as Figure 3 will allow foresters and forest managers to see precisely where longleaf previously thrived. This should aid in efforts to re-establish the longleaf resource.

Configuration, the other component of landscape diversity, is undergoing subtler changes. For the three physiographic provinces the pattern appears random, forest cover types occur in small (relative to the landscape) well intermixed patches. On the Piedmont there is more aggregation in 1972 and 1990 than in 1982, hence the low *RC* value in 1982.

A host of tools are today available to policy makers and forestry practitioners to assess the state of forests and to help guide in their management for sustainability. In dealing with landscape or regional scales of resolution, GIS and landscape indices are important tools for characterizing and comparing landscape diversity.

REFERENCES

Agee, J. K, and D. R. Johnson. 1988. Introduction to ecosystem management. *In* Agee, J. K., and D. R. Johnson, eds., Ecosystem management for parks and wilderness. Univ. of Washington Press, Seattle, WA:3-14.

Barnett, J. P., D. K. Lauer, and J. C. Brissette. 1990. Regenerating longleaf pine with artificial methods. *In* Farrar, Jr., R. M., ed., Proc. symp. on the management of longleaf pine. Gen. Tech. Rep. S0-75. U.S. Department of Agriculture, Forest Service, Southern Forest Experiment Station, New Orleans, LA:72-93.

Basharin, G. P. 1959. On a statistical estimate for the entropy of a sequence of independent random variables. *In* Artin, N., ed., Theory of probability and its applications, Vol. IV. (Translation of Teoriya Veroyatnostei I ee Pvimeneniya). Society for Industrial and Applied Mathematics, Philadelphia, PA:333-336.

Council on Environmental Quality. 1990. Linking ecosystems and biological diversity. *In* Environmental quality: the twenty-first annual report of the Council on Environmental Quality. U.S. Government Printing Office, Washington, DC: 135-187.

Environmental Systems Research Institute, Inc. 1992. ARC/INFO® user's guide: ARC command references. Environmental Systems Research Institute, Inc., Redlands, CA.

Hodgkins, E. J., M. S. Golden, and W. F. Miller. 1979. Forest habitat regions and types on a photomorphic-physiographic basis: a guide to forest site classification in Alabama-Mississippi. Southern Coop. Series No. 210. Ala. Agr. Exp. Sta., Auburn Univ., Auburn, AL. 64 p.

Li, H., and J. F. Reynolds. 1993. A new contagion index to quantify spatial patterns of landscapes. Landscape Ecology 8:155-162.

May, D. M. 1990. Stocking, forest type, and stand size class—the Southern Forest Inventory and Analysis unit's calculation of three important stand descriptors. Gen. Tech. Rep. S0-77. U.S. Department of Agriculture, Forest Service, Southern Forest Experiment Station, New Orleans. 7 p.

Naveh, L., and A. S. Lieberman. 1994. Landscape ecology: theory and application, 2nd ed. Springer-Verlag, New York. 360 p.

Oliver, C. D. 1992. Achieving and maintaining biodiversity and economic productivity: a landscape approach. J. For. 90(9):20-25.

O'Neill, R. V., J. R. Krummel, R. H. Gardner, G. Sugihara, B. Jackson, D. L. DeAngelis, B. T. Milne, M. G. Turner, B. Zygmont, S. W. Christensen, V. H. Dale, and R. L. Graham. 1988. Indices of landscape pattern. Landscape Ecology 1:153-162.

Pielou, E. C. 1975. Ecological diversity. John Wiley & Sons, New York. 165 p.

Salwasser, H. 1990. Conserving biological diversity: a perspective on scope and approaches. For. Ecol. Manage. 35:79-90.

Tickner, W. D. 1992. A vision for the future: an environmental scenario for the 21st century. J. For. 90(10):41-44.

Artificial Nest Predation Dynamics Along a Forest Fragmentation Gradient: A Preliminary Analysis

Erin Bayne

Keith Hobson

ABSTRACT. Studies in eastern North America suggest that nest predation on forest songbirds increases with habitat fragmentation. However, the majority of these studies have been conducted in highly fragmented suburban/rural deciduous forest habitat, making generalization of the results difficult. The objective of this study was to examine artificial nest predation dynamics along a fragmentation gradient (farm woodlots, logged forest stands and contiguous forest) in the conifer dominated Boreal Mixedwood. Predation was significantly higher in farm woodlots (78.8% edge and 78.5% interior) than the forest interior in contiguous and logged areas (42.8% and 41.9%, respectively). Predation at logged edges (60.0%) was not significantly different from either the woodlots or the forest interiors. All land-uses showed a variety of avian and mammalian predators. Using timer nests, we determined that predation was highest immediately after sunrise, with a second peak around sunset. Census results suggest that farm woodlots have significantly more red squirrels and a very different corvid community than forested areas. Overall, frag-

Erin Bayne is Graduate Student in the Department of Biology, University of Saskatchewan, Canada.

Keith Hobson is Researcher with the Canadian Wildlife Service, Saskatoon, Saskatchewan, Canada.

[Haworth co-indexing entry note]: "Artificial Nest Predation Dynamics Along a Forest Fragmentation Gradient: A Preliminary Analysis." Bayne, Erin, and Keith Hobson. Co-published simultaneously in *Journal of Sustainable Forestry* (Food Products Press, an imprint of The Haworth Press, Inc.) Vol. 5, No. 1/2, 1997, pp. 263-278; and: *Sustainable Forests: Global Challenges and Local Solutions* (ed: O. Thomas Bouman, and David G. Brand) Food Products Press, an imprint of The Haworth Press, Inc., 1997, pp. 263-278. Single or multiple copies of this article are available for a fee from The Haworth Document Delivery Service [1-800-342-9678, 9:00 a.m. - 5:00 p.m. (EST). E-mail address: getinfo@haworth.com].

263

mentation due to agriculture seems to have a far greater impact on nest predation in the Boreal Forest than does logging and is in part, due to changes in the predator community. Further work is required to confirm the relative importance of nest predators in different land-uses. Clarification of the impact that logged edges have is also required. *[Article copies available for a fee from The Haworth Document Delivery Service: 1-800-342-9678. E-mail address: getinfo@haworth.com]*

INTRODUCTION

A growing body of evidence suggests that forest songbirds, especially neotropical migrants (those species that winter in the Caribbean, Central America or South America but breed in temperate North America) are in decline (Whitcomb et al., 1981; Ambuel and Temple, 1982). The most commonly cited reason for these declines is increased habitat fragmentation (Askins et al., 1990; Hagan and Johnston, 1992). Fragmentation may impact forest songbirds in a variety of ways, including: (1) decreased forest area available as habitat; (2) smaller patches of habitat may be of poorer quality resulting in a reduction in the number of microhabitats available for nesting and foraging (Karr and Freemark, 1983; Lynch and Whigham, 1984; Douglas et al., 1992); (3) forest patches may be smaller than the minimum territory size for a particular species (Forman et al., 1976; Moore and Hooper, 1975); (4) small forest patches may have a greater probability of extinction due to stochastic changes in population size (Diamond, 1984); (5) changes in biotic interactions including increased predation, higher brood parasitism and greater competition from generalist species (Ambuel and Temple, 1983; Brittingham and Temple, 1983; Martin, 1988).

Of all the possible explanations for the decline of forest songbirds, most people agree that altered nest predation dynamics is likely one of the most important factors (Rudnicky and Hunter, 1993; Martin, 1993). Fragmentation can change nest predation rates by decreasing the amount of forested area and subsequently increasing the relative amount of edge. This may increase a predator's chance of finding a nest as a smaller area has to be searched, while edges may act as "ecological traps" where predators focus their activity (Gates and Gysel, 1978; Wilcove et al., 1986; Andren and Angelstam, 1988).

Habitat fragmentation has also significantly altered the distribution and abundance of nest predators relative to the conditions under which many bird species evolved (Martin, 1987; Clark and Nudds, 1991). Increased numbers of generalist predators such as crows (*Corvus branchrhyncos*), magpies (*Pica pica*) and skunks (*Memphitis memphitis*) are often observed at man-made edges in agricultural and urban landscapes (Andrén, 1992). Such increases are often thought to result from a more predictable food supply (e.g., grain left in fields, human refuse, roadkill, etc.) provided by human settlement. This is compounded by a lack of large predators in many human dominated landscapes, which historically controlled populations of smaller generalist predators.

Most research into habitat fragmentation and nest predation dynamics has been conducted in highly fragmented suburban/urban deciduous ecosystems in the eastern United States (Rudnicky and Hunter, 1993; Welsh and Healy, 1993). Studies of fragmentation, where the matrix remains forested (e.g., logging), are far less common and suggest that a number of trends observed in agricultural landscapes do not apply. "Edge-effects" are often not as severe in forested areas as they are in agricultural areas. Edges created by clear-cutting are often abrupt with little transition zone vegetation. Lack of brushy vegetation at clear-cut edges may not attract nesting birds relative to the forest interior. Without greater foraging reward, predators may not be attracted to edges (Angelstam, 1986; Small and Hunter, 1989). As well, forested areas usually lack generalist predators like crows, magpies and skunks. Instead, predators in forested areas tend to be animals such as red squirrels (*Tamiasciurus hudsonicus*), eastern chipmunks (*Tamias striatus*), deer mice (*Peromyscus maniculatus*), and ravens (*Corvus corax*) (Reitsma et al., 1990; Nour et al., 1993).

Generalizing the results of habitat fragmentation studies conducted in agricultural deciduous landscapes to coniferous dominated logged areas is inadequate due to the vast differences in habitats and potential predator species. The goal of this study was to determine if different levels of fragmentation (farm woodlots—severe fragmentation, logged areas—moderate fragmentation, contiguous forest—minor fragmentation) influenced predation rate on artificial nests in the southern Boreal Mixedwood forest.

METHODS

Study Sites

In 1994, nine forest stands in three different land-use areas in and around the Prince Albert Model Forest (53°50′N latitude; 105°50′W: hereafter referred to as model forest) were examined. The land-uses included: (1) Mature "forest" in Prince Albert National Park where greater than 60% of the forested land is 80 years or older and only non-consumptive recreation activities are allowed. (2) "Logged" sites along the Snowfield Road where logging activity has created a patchwork mosaic of different successional age-classes encased in a forest matrix. (3) "Farm" sites around Paddockwood, Saskatchewan where an area of once contiguous forest has been cleared for agriculture (grain and cattle production) over the last fifty years leaving mature woodlots (> 80 years old) isolated from other forested areas by a matrix of grain fields and pasture.

All study sites were mature (70+ years) mixedwood forest comprised mainly of white spruce (*Picea glauca*) and trembling aspen (*Populus tremuloides*). The "forest" sites were at least 100 m from any road or trail opening and were surrounded by mature forest completely. "Logged" sites were isolated on at least three sides by regenerating forest caused by logging that occurred within the last 15 years. "Farm" sites were completely isolated by agricultural land on all four sides (woodlots were 10, 22 and 39 ha).

Nest Layout

Artificial nests were commercially-manufactured open-cup nests (10 cm diameter by 6 cm depth) made of woven straw (e.g., Wilcove, 1985). Each nest contained one Japanese quail (*Coturnix japonica*) and one painted plasticine egg of the same size and colour. To avoid human scent, nests were left outdoors for 5 days before placement and quail eggs were washed with well water. After that nests and eggs were only handled while wearing rubber gloves and rubber boots. Scent was further reduced by placing vegetation from the sites in the nest. Nests were considered to be destroyed if any of the eggs were missing, broken, cracked or if any marks were left on the plasticine egg.

In two farm woodlots and two logged stands, edge and interior transects were established. Edge transects were 20 nests long with nests spaced 20 m apart, following the contour of the stand. Two interior transects of 10 nests each were located 100 m from any edge and at least 50 m apart. Interior nests were also spaced at 20 m intervals but were alternately placed 5 m away from the transect line to reduce linear distribution. In the forest area where no edge existed, the interior transects were also 50 m apart, but were 20 nests long.

Nests were placed alternately along the transects on the ground and in shrubs (1-2 m above the ground). At each nest, the percent of the nest visible from above, below and the four cardinal directions were estimated from a distance of 2 m to give an average visibility. An estimate of percent cover (0, 1, 5, 10, 20, 30% etc.) was calculated for conifers, deciduous trees, shrubs, grasses, herbs, moss and slash/dead material for 4 height strata (0-0.25, 0.25-1, 1-3, >3 m) in a $1m^2$ quadrant centered around each nest.

The experiment was conducted twice in 1994, with each trial lasting 12 days. Period 1 corresponded with the onset of breeding in forest songbirds at our latitude (June 6-June 18), while period 2 corresponded with nesting attempts made by late-nesting and re-nesting birds (July 2-July 14). Nest location was not the same between trials as different edge and interior transects were used.

Predator Identification and Temporal Dynamics

Each land-use had another site where nests were laid out in the same manner but each transect (edge/interior) included only 16 nests, each fitted with a digital timer. These timers were designed to record the time and date when predation events occurred. As this was the only difference between nests, these nests were also used in calculation of overall predation rate.

Twenty-four remote camera systems were used to photograph predators as they took eggs. The cameras were located along the same transects as the artificial nests. When preyed on, cameras were moved to an alternate location on the transect or to an area of high predation at another site. The number of camera nights was not equal in all land-uses.

Predators were also identified by teeth or beak marks left in plasticine eggs. Eggs destroyed by avian predators were not assigned to species due to considerable overlap in beak size. Mammalian predators were assigned to 4 broad categories based on dentition patterns: (1) shrew size animal; (2) mouse size animal; (3) squirrel size animal; (4) mammals larger than squirrels. The categories were based on measurements of teeth length and width from the skulls of museum specimens. Identification was aided by comparison of eggs collected from camera nests where the predator was identified.

Predator Abundance

The relative abundance of different corvid species in the three land-uses was estimated by means of a Breeding Bird Survey done between sunrise and 7 a.m. A 30 km section of road in each land-use was driven once in June and once in July. Every km, a 3 minute census was conducted where all corvids heard and seen were recorded. The mean number of each species was used in subsequent analysis.

The relative abundance of red squirrels was also estimated in all land-uses by means of a 10 minute acoustic census conducted between sunrise and 10 a.m. At each site, two to five listening stations, depending on the size and shape of the stand were established. These stations were 100 m from an edge and were at least 250 m apart. Each census was treated as independent for subsequent analysis.

A preliminary study to estimate trapping success of small mammals was conducted in late July and early August. Trapping was done in two sites in each land-use. Edge and interior trap transects were located in farm and logged areas while two interior transects were located in the forest area. Transects were 15 traps long with traps spaced by 10 m. Nine Sherman-style box traps and 6 #745 Havaheart squirrel traps were employed in a standardized pattern. Traps were set around 6 p.m. and were checked at 10 a.m. the next day. Traps were set for two day intervals in each land-use.

RESULTS

Nest Predation Rate

Overall, there was no difference in the percentage of nests destroyed between trials (arcsine transformed), so data for individual land-uses was pooled. There were significant differences in the percentage of nests destroyed per site between different treatments (F = 4.9155, P > 0.01). Tukey's HSD comparisons suggested that predation in the forest interior and logged interior were not significantly different (41.9% and 42.8%, respectively) but were significantly different from farm edge and farm interior (78.8% and 78.5%, respectively). Predation rate along the logged edge was not significantly different from other treatments (60.0%).

Identification of Predators Using Cameras

Thirty-five photographs of predation events were obtained. Red squirrels were the dominant predator in all land-uses (63% of all photographs; see Table 1). However, a variety of mammalian (86%) and avian predators (14%) were found in all other land-uses. Edges

TABLE 1. Predators identified by photographs in each land use

	Farm Edge	Farm Interior	Logged Edge	Logged Interior	Forest Interior
Crow	1	0	0	0	0
Magpie	2	0	0	0	0
Gray Jay	0	0	0	2	0
Red Squirrel	7	7	4	0	4
Deer Mouse	0	1	0	1	3
Striped Skunk	1	0	0	1	0
Porcupine	1	0	0	0	0
# Camera Nights	118	102	57	81	171

along farm woodlots had the highest predator richness (n = 5). Ground nests were almost exclusively destroyed by mammals (95.5%) while shrub nests were destroyed by birds (n = 4; 30.8%) and by red squirrels (n = 9; 69.2%). Considerably more pictures are required before statistical analysis between land-uses would be applicable.

Identification of Predators Using Plasticine Eggs

Of the 118 plasticine eggs recovered, 16.0% were destroyed by birds, 1.7% by large mammals, 39.0% by "mice," 2.5% by "shrews," 35.6% by "squirrels" and 5.9% could not be identified. These were not evenly distributed in all land-uses (see Table 2). Further, shrub nests were depredated almost exclusively by birds and squirrels. Predation due to mice was mainly on ground nests (see Table 2).

Temporal Dynamics

The time and date for 97 predation events was recorded. Predator activity was not distributed evenly throughout the day (Watson $U^2 =$ 0.347, P < 0.002: see Figure 1). Activity was bimodal with a peak

TABLE 2. Proportion (%) of plasticine eggs destroyed by five classes of predators in each treatment

	Farm Edge n = 18	Farm Interior n = 17	Logged Edge n = 26	Logged Interior n = 12	Forest Interior n = 45	Ground (Pooled) n = 74	Shrub (Pooled) n = 44
Bird	16.7	11.8	34.6	25.0	4.4	4.1	36.4
Shrew	0	0	7.69	0	2.2	2.7	2.7
Mouse	38.9	23.5	26.9	41.7	51.1	58.2	4.5
Squirrel	38.9	58.8	26.9	25.0	33.3	28.4	47.7
Large Mammal	0	0	0	0	2.2	2.7	0
Unidentified	5.6	5.9	3.9	8.3	6.7	4.1	9.1

centered early in the morning (172 minutes after sunrise ± 152 minutes standard deviation) and a second peak occurring in the evening (892 minutes after sunrise). Contrary to expected, very little predation occurred during the night (7.2%).

Nest Site Characteristics

No significant difference was found between predation rate on ground or shrub nests (62.2% vs. 60.5%, respectively; t = 0.024, P > 0.80; all land-uses pooled). Mann-Whitney U-tests were used to test whether there were any vegetative differences or visibility differences between destroyed and surviving nests. These were calculated for both ground and shrub nests. There was no significant difference between destroyed and undisturbed shrub nests for any measured variable. Ground nests at farm edges, logged edges and

FIGURE 1. Temporal distribution of predation events in one hour intervals relative to sunrise (all land-uses pooled). Each circle represents two occurrences.

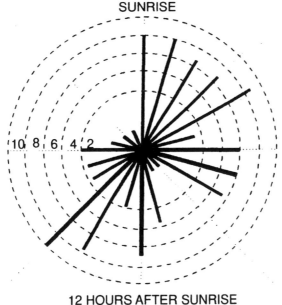

SUNRISE

10 8 6 4 2

12 HOURS AFTER SUNRISE

forest interior were more likely to survive if the grass cover was greater. The amount of deciduous cover in the logged interior was significantly higher at destroyed ground nests.

Red Squirrel Abundance

Red squirrels were significantly more abundant in farm woodlots (4.11 ± 1.71 squirrels detected per 10 minute census) than in the logged (1.42 ± 1.16) or forested areas (1.79 ± 1.27) (Kuskal Wallis $X^2 = 20.7228$, P < 0.001).

Corvid Abundance

The relative abundance of different corvid species was significantly different between land-uses ($X^2 = 171.32$, P < 0.001). However, there was no significant difference in total abundance between logged and forest land-uses for any species ($X^2 = 5.72$, P > 0.10). Crows and magpies were found almost exclusively in the farm area while gray jays and blue jays were found exclusively in forest and logged areas (see Table 3). Ravens were the only species that were found in all land-uses, but were far more abundant in the forest and logged areas. The relative density of birds was significantly higher in the farm (61 birds per 30 km; $X^2 = 31.13$, P < 0.001) than the forest or logged areas (26.5 vs. 16.0 birds per 30 km). There was no significant difference in the relative density of birds between forest and logged areas ($X^2 = 2.594$, P > 0.10).

TABLE 3. Mean number of corvids (± one S.D.) detected per land use

	Farm	Logged	Forest
Crow	47.0 ± 2.0	0	1.5 ± 0.71
Magpie	11.5 ± 0.71	0	0
Raven	2.5 ± 0.71	9.5 ± 2.12	14.5 ± 3.54
Gray Jay	0	6.0 ± 1.41	8.5 ± 0.71
Blue Jay	0	0.5 ± 0.71	2.0 ± 1.41

Small Mammal Abundance

The three most common species in all land-uses were deer mice (*Peromyscus maniculatus*; 55% of all captures: see Table 4), boreal red-backed voles (*Clethrionomys gapperi*: 31.1%) and red squirrels (*Tamasciurius hudsonicus*: 10.6%). Since other species had relatively low abundance, subsequent analysis was conducted on these three species individually. Since there was unequal trapping effort between land-uses, all analysis was conducted based on captures per 120 trap nights (TN).

Relative abundance of red squirrels was significantly different between land-uses ($X^2 = 21.7977$, $P < 0.0001$) with the highest relative abundance in farm woodlot interiors (11 per 120 TN). When farm woodlot interiors were removed from the analysis there was no significant difference ($X^2 = 4.399$, $P > 0.10$). Relative abundance of red-backed voles was not significantly different between land-uses ($X^2 = 6.033$, $P > 0.10$).

The relative abundance of deer mice was significantly different between land-uses ($X^2 = 10.641$, $P < 0.05$) with highest abundance in the forest interior (22.667 per 120 TN). Logged edge was not included in the analysis as no deer mice were caught here. When forest was removed from the analysis there was no significant difference between land-uses in the relative abundance of deer mice ($X^2 = 2.744$, $P > 0.25$).

TABLE 4. Number of small mammals per 120 trap nights caught in each land-use and edge type

	Farm Edge	Farm Interior	Logged Edge	Logged Interior	Forest Interior
Red Squirrel	4	11	1	1	1
Boreal Red-backed Vole	6	3	5	8	11
Deer Mouse	12	13	0	6	22

DISCUSSION

In this study, predation rate was highest in the farm woodlots. However, unlike many studies in agricultural areas (Gates and Gysel, 1978; Wilcove et al., 1986; Møller, 1989) no edge-effect was observed. Edge-effects are often associated with the invasion of generalist predators from the surrounding matrix (Møller, 1989; Andren, 1992). A distinct increase in the abundance of generalist corvids (crows and magpies) was observed in the farm areas. Further, photographic evidence suggested that generalist predators did occur at farm woodlot edges. However, there was not an increased predation rate when compared to the interior suggesting that the importance of generalist predators may not be as significant as predators residing in forest patches. Photographic evidence suggested that the increased predation rate in farm woodlots, especially in the interior, seems to be caused by red squirrels. This concurs with trapping and census data which found red squirrels to be far more abundant in the interior of farm woodlots than at the edge of woodlots or in the logged and forest areas.

The forest interior and logged interior had the lowest rates of predation and were not significantly different. Predation rate at the logged edge was higher than the logged or forest interiors and lower than the farm woodlots, but not significantly so. This result must be viewed with caution as there was a significant difference between trials. Most studies done in forested landscapes have not found any edge effect (e.g., Ratti and Reese, 1988; Storch, 1991). The presence of an edge-effect in this study does not seem to be due to the invasion of generalist predators such as crows (e.g., Yahner and Scott, 1988) as they were not present in this land-use. Further, logged edges had relatively low abundance of small mammals, including red squirrels. Possibly, logged edges act as travel lanes for predators such as birds (Chasko and Gates, 1982). Further work is required to determine if the edge-effect observed in 1994 is significant, and if so what predators are causing it.

Micro-site differences (e.g., nest location and vegetative composition) did not have a significant role in determining the fate of nests. Ground nests were destroyed as often as shrub nests, which is in contrast to other studies which suggest that ground nests are destroyed more frequently (Wilcove, 1985; Gibbs, 1991). However, this trend is far from definitive and Martin (1993) has argued that in forested habitats this trend is reversed. Likely, the result depends on the local predator community. Photographic evidence suggests that mammals are far more likely to be the predators on ground nests (all predation by large mammals and deer mice was on ground nests). Of the 5 photographs of birds, 4 were taken in shrubs suggesting that these nests may be more susceptible to avian predation. Squirrels depredated similar numbers of ground and shrub nests (13 and 9, respectively).

The only vegetative variable that seemed to predict predation rate was grass cover. Ground nests with greater grass cover suffered reduced predation likely due to reduced visibility. However, visibility was not significantly different between destroyed and undisturbed nests. Perhaps, nests with a thick grass cover are less likely to be discovered by predators due to reduced mobility in thick grass.

The temporal dynamics of predation were contrary to expected, based on the types of predators responsible for predation events. Mammalian predation has long been assumed to occur at night while avian predation is thought to occur during the day. In a camera study by Picman and Schrmil (1994), predation due to birds occurred exclusively during day-light hours while mammalian activity was highest between 16:00 and 6:00 hours with very little occurring near dawn. In this study, peak predation was seen around dawn with a second peak occurring near twilight. However, skunks and raccoon were the dominant mammalian predators in Picman and Schrmil's study, and these predators likely have different activity patterns than the dominant predator in this study which seemed to be the red squirrel. Further work using cameras and timers together is required to better establish the temporal patterns of predators in the boreal forest.

It is widely accepted that artificial nests only provide a relative index of predation rate, although many studies have shown that they suffer similar rates of predation as natural nests (Martin 1987; Yahner and Voytko 1989; Major 1990; Yahner 1991). A major concern that has been expressed is the idea of trap-lining, where predators determine that layout of artificial nests and move quickly from one to the other. The timer data suggests that such a pattern did not occur in this study, as nests were never destroyed immediately after one another. Another concern is that predators watch researchers when nests are being placed, and destroy the nests soon after the observer has left. The majority of destroyed timer nests were destroyed early in the experiment (17% of all nests destroyed were destroyed on the day of placement). However, this did not occur immediately after placement but seemed to occur the following evening.

The decline of forest songbirds has been linked to changes in predator community that occur with fragmentation. The results of this study suggest that farm woodlots do have much higher predation rates than logged or forested landscapes and this is partially due to difference in predator community. Obviously, extrapolating the results of studies conducted in farmland to logged landscapes is inadequate. More work is required to determine how different land-uses influence predator communities, especially red squirrels.

REFERENCES

Ambuel, B., and S. A. Temple. 1982. Songbird populations in southern Wisconsin: 1954 and 1979. J. Field Ornithol. 53:149-158.

Andren, H. 1992. Corvid density and nest predation in relation to forest fragmentation: a landscape perspective. Ecology 73:794-804.

Andren, H., and P. Angelstam. 1988. Elevated predation rates as a edge effect in habitat islands: experimental evidence. Ecology 69:544-547.

Angelstam, P. 1986. Predation on ground-nesting birds' nests in relation to predator densities and habitat edge. Oikos 47:365-373.

Askins, R. A., J. F. Lynch, and R. Greenberg. 1990. Population declines in migratory birds in eastern North America Curr. Ornithol. 7:1-57.

Chasko, G. G., and J. E. Gates. 1982. Avian habitat suitability along a transmission-line corridor in an oak-hickory forest region. Wildl. Monogr. 82:1-41.

Clark, R. G., and T. D. Nudds. 1991. Habitat patch size and duck nesting success: the crucial experiments have not been performed. Wildl. Soc. Bull. 19:534-543.

Diamond, J. M. 1984. 'Normal' extinctions in isolated populations. Pages 191-246 in M. H. Nitecki, ed. Extinctions. University of Chicago Press, Chicago, Illinios, USA.

Douglas, D. C. , Ratti, J. T., Black, R. A., and J. R. Alldredge. 1992. Avain habitat associations in riparian zones of Idaho's centennial mountains. Wilson Bull. 104:485-500.

Forman, R. T. T., A. E. Galli, and C. F. Leck. 1976. Forest size and avian diveristy: New Jersey woodlots with some land use implications. Oecologia 26:1-8.

Freemark, K., and B. Collins. 1992. Landscape ecology of birds breeding in temperate forest fragments. Pages 443-454 in J. M. Hagan III and D. W.

Johnston. eds. Ecology and conservation of neotropical migrant landbirds. Smithsonian Inst. Press, Washington, D.C.

Gates, J. E., and L. W. Gysel. 1978. Avian nest dispersion and fledgling success in field-forest ecotones. Ecology 59:871-883.

Gibbs, J. P. 1991. Avian nest predation in tropical wet forest: an experimental study. Oikos 60:155-161.

Hagan, J. M. III, and D. W. Johnston, eds. 1992. Ecology and conservation of neotropical migrant landbirds. Smithson. Inst. Press, Washington, D.C., USA.

Karr, J. R. and K. E. Freemark. 1983. Habitat selection and environamental gradients: dynamics in the "stable" tropics. Ecoolgy 64:1481-1494.

Lynch, J. F. and D. F. Whigham. 1984. Effects of forest fragmentation on breeding bird communities in Maryland. Biol. Conserv. 34:333-352.

Major, R. E. 1990. The effect of human observers on the intensity of nest predation. Ibis 132:608-612.

Martin, T. E. 1987. Artificial nest experiments: effects of nest appearance and type of predator. Condor 89:925-928.

Martin, T. E. 1993. Nest predation among vegetation layers and habitat types: revising the dogmas. Am. Nat. 141:897-913.

Moller, A. P. 1989. Nest site selection across field-woodland ecotones: the effect of nest predation. Oikos 56:240-246.

Moore, N. W., and M. D. Hooper. 1975. On the number of bird species in British woods. Biol. Conserv. 8:239-250.

Nour, N., E. Matthysen, and A. A. Dhondt. 1993. Artificial nest predation and habitat fragmentation: different trends in bird and mammal predators. Ecography 16:111-116.

Picman, J., and L. M. Schiriml. 1994. A camera study of temporal patterns of nest predation in different habitats. Wilson Bull 106:456-465.

Ratti, J. T., and K. P. Reese. 1988. Preliminary test of the ecological trap hypothesis. J. Wildl. Manage. 52:484-491.

Reitsma, L. K, R. T. Holmes. and T. W. Sherry. 1990. Effects of removal of red squirrels, *Tamiasciurus hudsonicus*, and eastern chipmunks, *Tamias striatus*, on nest predation in a northern hardwood forest: an artificial nest experiment. Oikos 57:375-380.

Rudnicky, T. C., and M. L. Hunter. 1993. Avian nest predation in clearcuts, forests, and edges in a forest-dominated landscape. J. Wildl. Manage. 57:358-364.

Small, M. F., and M. L. Hunter. 1988. Forest fragmentation and avian nest predation in forested landscapes. Oecologia 76:62-64.

Storch, I. 1991. Habitat fragmentation, nest site selection, and nest predation risk in Capercaillie. Ornis Scan. 22:213-217.

Welsh, C. J. E., and W. M. Healy. 1993. Effect of even-aged timber management on bird species diversity and composition in northern hardwoods of New Hampshire. Wildl. Soc. Bull. 21:143-154.

Whitcomb, R. F., D. S. Robbins, J. F. Lynch, B. L. Bystrak, M. K. Klimkiewitz, and D. Bystrak. 1981. Effects of forest fragmentation on avifaunas of the eastern deciduous forest. Pages 125-205 in R. L. Burgess and D. M. Sharpe eds. Forest island dynamics in man-dominated landscapes. Springer-Verlag, New York, New York, USA.

Wilcove, D. S. 1985. Nest predation in forest tracts and the decline of migratory songbirds. Ecology 66:1211-1214.

Wilcove, D. S., C. H. McLellan, and A. P. Dobson. 1986. Pages 237-256 in M. E. Soule ed. Conservation biology: the science of scarcity and diversity. Sinauer Associates. Sunderland, Mass.

Yahner, R. H. 1991. Avian nesting ecology in small even-aged aspen stands. J. Wildl. Manage. 55:155-159.

Yahner, R. H., and D. P. Scott. 1988. Effects of forest fragmentation on depredation of artificial nests. J. Wildl. Manage. 52:158-161.

Yahner, R. H., and R. A. Voytko. 1989. Effects of nest-site selection on depredation of artificial nests. J. Wildl. Manage. 53:21-25.

Using Micropropagation
to Conserve Threatened Rare Species
in Sustainable Forests

J. L. Edson
D. L. Wenny
A. D. Leege-Brusven
R. L. Everett

ABSTRACT. For forests to be sustainable, viable populations of rare plants should be maintained. Where habitat management alone cannot conserve species threatened by human activity, micropropagation may advance species recovery. Micropropagation protocols were developed for Pacific Northwest endemics; *Hackelia venusta*, *Douglasia idahoensis*, *Astragalus* species, and *Cornus nuttallii*. Microshoots and seed were multiplied and rooted on nutrient media containing minimal levels of cytokinin and auxin growth regulators to maintain stable gene expression in plantlets. Acclimatized plantlets were reintroduced to protected habitat or propagated for further environmental experiments. Micropropagation serves a useful off-site role in sustaining Pacific Northwest forests by maintaining viability of certain threatened rare plants. *[Article copies available for a fee from The Haworth Document Delivery Service: 1-800-342-9678. E-mail address: getinfo@haworth.com]*

J. L. Edson, D. L. Wenny, A. D. Leege-Brusven are affiliated with the Department of Forest Resources, University of Idaho, Moscow, ID, USA.

R. L. Everett is affiliated with the USDA Forest Service Forestry Sciences Laboratory, Wenatchee, WA, USA.

[Haworth co-indexing entry note]: "Using Micropropagation to Conserve Threatened Rare Species in Sustainable Forests." Edson, J. L. et al. Co-published simultaneously in *Journal of Sustainable Forestry* (Food Products Press, an imprint of The Haworth Press, Inc.) Vol. 5, No. 1/2, 1997, pp. 279-291; and: *Sustainable Forests: Global Challenges and Local Solutions* (ed: O. Thomas Bouman, and David G. Brand) Food Products Press, an imprint of The Haworth Press, Inc., 1997, pp. 279-291. Single or multiple copies of this article are available for a fee from The Haworth Document Delivery Service [1-800-342-9678, 9:00 a.m. - 5:00 p.m. (EST). E-mail address: getinfo@haworth.com].

279

INTRODUCTION

For forests to be sustainable, viable populations of rare plants should be maintained. Many rare woody and herbaceous plants have tangible or potential value to mankind. Increased human activity, however, has altered or eliminated habitat, overharvested plant resources, introduced competitive or pathogenic organisms, or modified the atmosphere to threaten rare endemic forest species and small disjunct populations. Habitat management alone may be unable to prevent extinctions or genetic losses where taxa exist as few meta-populations, occupy narrow ecological niches, or fail to regenerate vigorously (Falk 1992). Off-site conservation can advance species recovery by increasing plant numbers and by reintroducing plant populations into protected habitat (Maunder 1992). As a technology for off-site conservation, micropropagation is fast, uses small amounts of seed or shoots, and may succeed when other methods fail (Fay 1992). Study objectives were to advance, through development and application of micropropagation technique, the recovery of threatened taxa endemic to the forests and rangelands of the Pacific Northwest.

MATERIALS AND METHODS

Hackelia venusta, Douglasia idahoensis, Astragalus species, and *Cornus nuttallii* were chosen for micropropagation based on perceived threats to their continued existence. Small amounts of seed were collected from abundant sources, but shoot tips were propagated where seed was scarce.

General Micropropagation Technique

Micropropagation involves surface sterilizing seed shoots or buds, inducing shoot growth in sterile culture, multiplying and elongating microshoots, rooting elongated shoots, and acclimatizing plantlets (Figure 1). Seeds or shoots were incubated on a gel containing Murashige and Skoog (MS) nutrient medium (Murashige and Skoog 1962) supplemented with cytokinin or auxin growth regulators to enhance shoot and root formation, respectively.

FIGURE 1. Schematic showing procedure for multiplication of microshoots in sterile culture

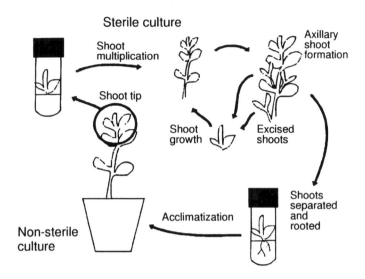

Producing True-to-Type Plants

In rare plant recovery, plantlets should be propagated from a comprehensive sampling of the genome and changes in genes or gene expression during culture should be avoided. High concentrations of growth regulators, long culture time, and adventitious shoots produced from callus may result in off-type plants (George 1993), whereas low levels of growth regulators and short culture time promote phenotypically stable axillary growth.

To find low levels of growth regulator adequate for propagation of axillary shoots, microshoots were initially multiplied on MS media containing benzyladenine (BA) from 0 to 10 micromoles. Multiplication rates were selected from plots of BA dose versus microshoot production so that required numbers of plantlets to establish new populations could be produced in minimal time. Minimal BA dosages corresponding to the chosen multiplication rates were predicted by inverse regression (Seber 1977). Microshoots were rooted without auxin treatment where possible or low levels of indoleacetic acid (IAA) where necessary. Plantlets (rooted micro-shoots) were transplanted to peat-perlite mixtures in a fog chamber (90% relative humidity) for several weeks and acclimatized to lower humidity and higher light intensity during a season's growth in greenhouse conditions before reintroduction.

APPLICATIONS AND DISCUSSIONS

A Highly Endangered Species

Hackelia venusta or showy stickseed (Boraginaceae), an herbaceous biennial endemic to the Washington Cascades (Hitchcock and Cronquist 1973), comprises a population of fewer than 100 individuals threatened by human activity in its roadside habitat. Because of seed scarcity, 1 shoot from each of 7 plants was multiplied (Figure 2) to develop a BA dose-response curve (Figure 3) for the species. From the plot, we estimated that 3 or fewer axillary shoots could be produced monthly by a low (<0.10 M) level of BA. Because 4 populations were to be reintroduced (1 ramet/genotype/population), 2 shoots/genotype/month needed to be produced in a minimum of 2 months. Inverse regression predicted minimal dosage of 0.02 MBA would produce 2 shoots/month.

A further on-site sample of 30 genotypes was randomly selected as representative of variation in the population. Microshoots of all 30 genotypes were multiplied (with 0.02 M BA) and rooted at 90-100% on a medium with low dosage of 0.5 M IAA. Plantlets grew vigorously in the greenhouse (Figure 4). First reintroductions resulted in survival of up to 95% of the plantlets (Figure 5) after 9 months on the new forest site. In addition to ongoing plant recovery actions, propagated clones are used to study taxonomy and reproductive biology of the species.

Vulnerability to Climate Change

Douglasia idahoensis (Primulaceae) is a showy herbaceous Idaho endemic (Henderson 1981) found in 24 widely distributed populations (some with fewer than 50 individuals). Climate warming could result in rapid decline of this pre-pleistocene relict confined to subalpine peaks and lacking upslope refugia.

FIGURE 2. Microshoot of *Hackelia venusta* on multiplication MS medium containing BA

FIGURE 3. Effect of BA concentration on microshoot multiplication in *H. venusta* after 1 month. Horizontal arrow represents microshoots per genotype to be produced and vertical arrow shows the predicted BA dose used

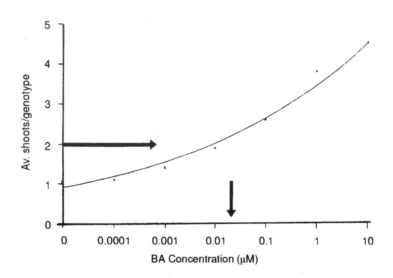

Microshoots were multiplied on MS medium with 0.04 M BA, and 60% rooting occurred without auxin supplement. Greenhouse and field growth has remained vigorous (Figure 6) for 2 years. We proposed trial plantings be initiated at higher altitude or latitude than current habitat to test the hypothesis that species viability would be enhanced in cooler refugia.

Micropropagating Seed

Micropropagation can increase survival of germinants in often critically small lots of seed collected from rare plants since seed germinated aseptically in culture avoids dampoff and other greenhouse pathogens.

An estimated 156 threatened species of *Astragalus* (Fabaceae) grow in the United States including *A. columbianus*, *A. amblytropis*, and *A. mulfordiae* (Sauer et al. 1979; Falk 1992). A single protocol using MS medium with 0.01 MBA successfully multiplied clones of the 3 species from sterile seedlings (Figure 7). More than 90% of

FIGURE 4. *H. venusta* plantlets (rooted microshoots) transplanted to the greenhouse

FIGURE 5. *H. venusta* plantlets reintroduced to a forest site

FIGURE 6. *Douglasia idahoensis* after 1 year of greenhouse growth

FIGURE 7. Sterile germinant of *Astragalus columbianus*

A. columbianus and *A. amblytropis* transplants to greenhouse and field survived and produced viable seed. Adaptability of *A. columbianus* is being tested in climate change experiments.

Unique Disjunct Population

Idaho's Pacific dogwood (*Cornus nuttallii*) is a once abundant, but now rare, unique coastal disjunct in the Northern Rockies. The trees, infected with dogwood anthracnose (*Discula destructiva*) produce little seed. Shoots were multiplied and rooted from initially disease-free greenhouse seedlings (Edson et al. 1994). Since transplanted plantlets became infected in the greenhouse after several years, reintroduction is not presently feasible. Present conservation actions are to maintain germplasm as seeds in seed banks at $-20\,^\circ$C and as microshoot cultures in gene banks with growth suspended at $1\,^\circ$C (Figure 8) until propagation of resistant genotypes allows reintroduction.

CONCLUSIONS

Micropropagation, as an off-site option, can help maintain biodiversity in sustainably managed forests of the Pacific Northwest where threatened taxa cannot be adequately protected by habitat management alone. Our methodology using low levels of growth regulators and short culture times should be useful to preserve phenotypic stability and advance plant recovery programs when propagation is necessary and feasible.

FIGURE 8. Microshoots of *Cornus nuttallii*

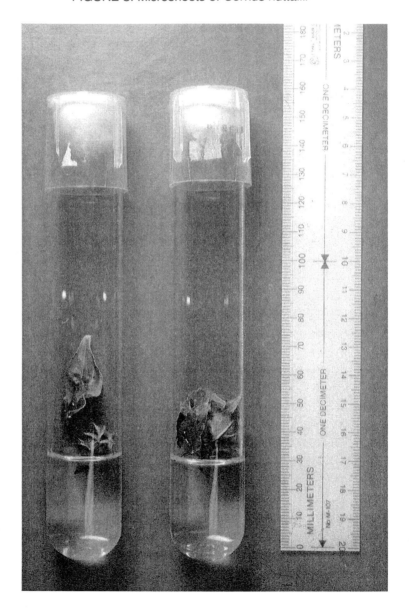

REFERENCES

Edson, J. L., D. L. Wenny, and A. Leege-Brusven. 1994. Micropropagation of Pacific Dogwood. HortScience 29(11): 1355-56.

Falk, D. A. 1992. From conservation biology to conservation practice: strategies for protecting plant diversity. In: Conservation biology. P.L.

Fay, M. F. 1992. Conservation of rare and endangered plants using in vitro methods. In Vitro Cell. Dev. Biol. 28P: 1-4.

Fiedler and S. K. Jain (ed.). 16:397-431. Chapman and Hall, New York.

George, E. F. 1993. Plant propagation by tissue culture. Part 1. The Technology. Exegetics Ltd. Great Britain.

Henderson, D. M. 1981. A new Douglasia (Primulaceae) from Idaho. Brittonia, 33(1): 52-56.

Hitchcock, C. L. and A. Cronquist. 1973. Flora of the Pacific Northwest. University of Washington Press, Seattle, Washington.

Maunder, M. 1992. Plant reintroduction: an overview. Biodiversity and conservation. 1:51-61.

Murashige, T. and F. Skoog. 1962. A revised medium for rapid growth and bioassays with tobacco tissue cultures. Physiol. Plant. 15:473-97.

Sauer, R. H., J. D. Mastrogiuseppe, and R. H. Smookler. 1979. *Astragalus columbianus* (Leguminosae)-rediscovery of an "extinct" species. Brittonia. 31(2):261-264.

Seber, G. A. F. 1977. Linear regression analysis. John Wiley & Sons, New York.

Microbial Inoculants
for Sustainable Forests

M. S. Reddy
L. M. Funk
D. C. Covert
D. N. He
E. A. Pedersen

ABSTRACT. Fungal root pathogens are widespread and may cause substantial seedling losses in conifer nurseries. Furthermore, poor seedling survival and growth on reforestation sites results in reduced forest regeneration. Use of microbial inoculants for disease control and plant growth promotion has become an important endeavour. A microbial culture collection of 500 strains was assessed for biological control of fungal root pathogens and/or plant growth promotion of conifer seedlings. Seven of these strains showed significant suppressive effects on various soil-borne fungal pathogens. On Douglas fir, two strains, RAL3 and 64-3, reduced disease caused by *Fusarium* by 7-42% in repeated growth room assays. The same strains significantly increased healthy stand of white spruce seedlings inoculated with *Fusarium* and *Pythium* in a conifer nursery, and increased the survival of bare-root white spruce seedlings planted on a reforestation site by 19-23%. Both strains also significantly increased new root and

M. S. Reddy, L. M. Funk, D. C. Covert, D. N. He and E. A. Pedersen are affiliated with Agrium Inc., AgBiologicals, Saskatoon, Saskatchewan, Canada.

[Haworth co-indexing entry note]: "Microbial Inoculants for Sustainable Forests." Reddy, M. S. et al. Co-published simultaneously in *Journal of Sustainable Forestry* (Food Products Press, an imprint of The Haworth Press, Inc.) Vol. 5, No. 1/2, 1997, pp. 293-306; and: *Sustainable Forests: Global Challenges and Local Solutions* (ed: O. Thomas Bouman, and David G. Brand) Food Products Press, an imprint of The Haworth Press, Inc., 1997, pp. 293-306. Single or multiple copies of this article are available for a fee from The Haworth Document Delivery Service [1-800-342-9678, 9:00 a.m. - 5:00 p.m. (EST). E-mail address: getinfo@haworth.com].

total plant dry weights. Strain RAL3 in commercial formulation maintained a viable population of about log 8-9 cfu/ml for over a year when stored at 5°C. Strain survival on seed varied with conifer species. No decreases in bacterial populations were observed on seeds of jack pine or Douglas fir after 37 to 44 days storage at 5°C, but decreases were observed on seeds of white spruce and Scots pine. This study has provided candidate beneficial microbial inoculants which offer promise for development of commercial inoculants for the forestry industry. *[Article copies available for a fee from The Haworth Document Delivery Service: 1-800-342-9678. E-mail address: getinfo@ haworth.com]*

INTRODUCTION

Forestry is an extremely important industry in many countries, including Canada (Reddy et al. 1993). With an increasing demand for forest products, foresters have turned toward more intensive management practices to increase productivity of forest lands. These methods include the breeding of forest trees with increased growth and superior wood characteristics, artificial regeneration using seedlings, control of competing vegetation, thinning stands to reduce competition between trees, fertilization to stimulate growth, and improved methods for harvesting and utilization of wood. Most of these practices are well understood and are used in modern forestry. Perhaps the most effective way of increasing productivity is the use of genetically improved material as planting stock. Unfortunately, this material is usually in short supply.

Over seven hundred million conifer seedlings are planted annually in Canada, making silviculture an extremely important industry. It is imperative that seedling quality be at a high level to allow for successful reforestation of harvested land. Seedling losses occur in conifer nurseries as well as on reforestation sites. Fungal pathogens inciting seed and/or root diseases in nurseries include *Fusarium, Cylindrocarpon, Cylindrocladium* and *Pythium*. Diseases caused by these pathogens may be important limiting factors in production of high quality seedlings in forest and conservation nurseries. All nurseries experience some losses to damping-off and root rot diseases, despite the best efforts at control. These losses may occur in several forms, the most obvious being dead and dying seedlings observed in nursery beds. The economic loss represented by this type of seedling mortality may vary with the age of the affected seedlings. Diseases may also damage seedlings, making them unsuitable for planting. Damaged seedlings are thrown away (culled) during lifting and packaging. Some diseased seedlings may escape culling or remain symptomless at the time of lifting. In these cases, pests are transported to field plantings, where they continue to cause economic losses by killing, stunting, or deforming transplanted seedlings.

Chemical pesticides were initially formulated for effectiveness on many soil-borne pathogens. This broad spectrum efficacy often resulted in destruction of both beneficial and injurious organisms (Baker and Cook 1974). However, resistance to these chemicals can develop rapidly in pathogens. In recent years, problems with pesticide resistance, toxicity to non-target organisms, and environmental contamination have greatly reduced the desirability of chemical fungicides (Campbell 1989). Recent public and government involvement in banning chemicals used in agriculture and forestry will undoubtedly make the use of chemical fungicides difficult at best. Therefore, foresters and nursery managers need to examine all alternatives for controlling fungal diseases. One of the most acceptable approaches to disease control is the use of naturally occurring microbial inoculants to reduce or suppress the activity of fungal pathogens (Reddy 1991).

Losses in forest productivity include poor seedling establishment and survival on reforestation sites due to factors other than disease. For example, root growth of transplanted seedlings is often limited, contributing to poor seedling health. Root system morphology is a major determinant of seedling success in the field. The goal of bare-root nursery managers is to produce high quality seedlings which can tolerate lifting, handling, and planting processes, and not only survive but grow competitively in the field. This goal is a challenging one to attain. No two nurseries are alike and within-nursery variation in soil and microclimate may be as great as that among nurseries. Seedling grading has been controversial because no scientifically based procedure has been developed for identifying which seedlings in a nursery will be the most competitive in the field. The economic impact due to seedling losses on reforestation sites, regardless of the cause, is substantial since the approximate cost to plant a single seedling is $1.00. Many diverse groups of bacteria commonly inhabit nursery soil. Several species are antagonistic toward common soil-borne pathogens (Reddy 1991; Reddy et al., 1991, 1992, 1993, 1994). In our program beneficial microorganisms specifically selected for forestry are being evaluated for: (1) Biological control of fungal root pathogens in conifer nurseries, (2) Enhancement of conifer seedling growth and survival on reforestation sites.

The purpose of this report is to draw the attention of forest managers and researchers to the potential commercial value of incorporating microbials as seed or root inoculants to increase productivity in intensive forestry programs. Selection of these bacteria was based upon availability, ease of manipulation, wide geographic and host range, and demonstrated benefits to a wide variety of host trees.

MATERIALS AND METHODS

Seed and seedlings used for experimentation purposes were obtained from commercial nursery suppliers and industry seed orchards. The best quality seed and seedlings, both genetically and physically, were selected.

Biological Control

Approximately 500 microbial strains isolated from conifer seedling rhizospheres, mycelial baits, and disease escape plants, and beneficial strains from other plant/pathogen systems were screened for suppression of Fusarium and Pythium diseases on Douglas fir and white spruce seedlings. Conifer seed was soaked in bacterial cell suspension in liquid proprietary formulation (log 3-5 cfu/seed). Seed was then planted in Magenta GA7 jars containing peat/vermiculite planting mix fertilized with a liquid nursery solution to 79% moisture (planting mix pH 5.2). The planting mix surface was inoculated with conidia of *F. oxysporum* in 1.0 mL spots under seeds (log 5 cfu/mL). Seeds were covered with granite grit and jars were capped with lids, randomized and incubated in a growth room for 4 weeks under an 18 hr photoperiod using a moderate heat stress (28-30°C) to enhance disease development (8 seeds/jar, 8 jars/treatment; *Fusarium* control treatment utilized 24 jars). A similar experiment was conducted in a commercial nursery at Prince Albert, Saskatchewan. At the nursery, treatments were replicated 10 times and were arranged in a randomized complete block design. Percentage germination and healthy stand were assessed during the experiment. All strains which resulted in a reduction in disease were tested two to four times. The data were analyzed using ANOVA and Fisher's protected LSD test.

Plant Growth Promotion Reforestation Trials

Select bacterial strains *Burkholderia cepacia* strain RAL3 and *Pseudomonas fluorescens* strain 64-3 were tested for enhancement of seedling growth and survival on field sites in British Columbia, Alberta, and Saskatchewan. The roots of container or bare-root white spruce seedlings were soaked in a bacterial suspension in liquid formulation (log 7 cfu/mL) for 15 to 30 min and planted in reforestation field trials in a randomized complete block design. Control seedlings received no bacteria. There were 10 replicates with 20 seedlings per treatment. Experimental plots were flagged in each corner with surveyors' flags, and wooden stakes were labelled with plot numbers and placed at the centre-front of each plot to allow identification during planting and assessment. Seedling survival, height (from ground line to the tip of the terminal bud), and root collar diameter were assessed at 4 and 15 months after planting. Dry weight (biomass production) of above ground parts (foliage, stem, and branches) of the seedlings and new roots were also measured. The results were analyzed using ANOVA and Fisher's protected LSD test.

Shelf-Life of Bacteria in Formulation and on Conifer Seed

The shelf-life of strains RAL3 and 64-3 were evaluated in proprietary commercial liquid formulation. The inoculum was produced in lab-scale fermenters and stored in commercial packages at 5°C and 20°C (room temperature). Shelf-life was assessed using standard serial dilution plate technique onto replicated Pseudomonas Agar F plates at zero time and at 2 month intervals for one year. Plates were incubated for 48 h at 30°C, colonies were counted, and the data were converted to colony forming units (cfu)/mL.

The shelf-life of strains RAL3 and 64-3 were also assessed on seed of white spruce, Douglas fir, jack pine, and Scots pine. Seeds were treated with rifampicin (rif) marked strains as a seed soak (0.3 mL/g seed) and stored at 5°C. There were 3 replications per treatment. Seeds were sampled to determine survival of the bacteria at various times up to 52 days after treatment. The sampling procedure was the same as described above. Shelf-life data were converted to cfu/seed.

RESULTS

The locations of experimental sites are shown in Table 1. Some of the microbial inoculants included in greenhouse bioassays, commercial nurseries, and on reforestation sites are shown in Table 2. For the sake of brevity, only results for two bacterial strains RAL3 and 64-3 are reported here.

TABLE 1. Field and greenhouse experimental sites, 1992-1994

Location	Date Started	Type
Duck Lake, SK	May, 1992	field
Vancouver, BC	July, 1992	greenhouse, field
Prince Albert, SK	April, 1993	greenhouse
Prince George, BC	June, 1993	field
Tappen, BC	June, 1993	field
Lac La Biche, AB	June, 1994	field

TABLE 2. Bacterial strains tested as microbial inoculants under laboratory, greenhouse, and field conditions

Strain Number	Strain Identification	Source of Isolation
17-29	*Pseudomonas putida*	northern soil
64-3	*Pseudomonas fluorescens*	canola rhizosphere
RAL3	*Burkholderia cepacia*	soybean rhizosphere
D31-3A	*Pseudomonas chlororaphis*	forest soil
D32-4D	*Pseudomonas* spp.	forest soil
D31-8B	*Pseudomonas chlororaphis*	forest soil
QP5	*Pseudomonas fluorescens*	peat bog

Biological Control

Forty-one bacterial strains reduced Fusarium disease in one or more biocontrol assays, but the data for many of the strains were highly variable. Strain 64-3 reduced the incidence of diseased Douglas fir seedlings compared to the non-treated control by 36 to 42% in two of four assays (data from fourth assay not presented) (Figure 1). Similarly, strain RAL3 reduced disease incidence by 24 to 26% in two of three assays; however, these reductions were not statistically significant. Both strains increased the healthy stand of white spruce seedlings infested with *Pythium* and *Fusarium* at a conifer nursery in Saskatchewan (Figure 2). In many cases the two strains also minimized symptom expression on roots infected with either pathogen. The importance of this result is currently being assessed in nursery trials.

FIGURE 1. Reduction of the incidence of Fusarium disease on Douglas fir seedlings by strains RAL3 and 64-3 in magenta jar assays. Asterisks denote significant (*p < 0.10, **p < 0.05) reduction in number of diseased seedlings compared to non-treated control, according to Dunnett's test.

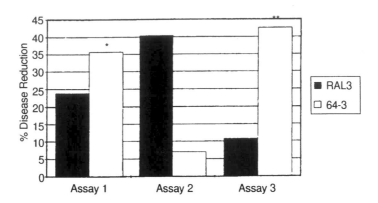

FIGURE 2. Increased healthy stand of white spruce seedlings infested with *Pythium* and *Fusarium* at a commercial conifer nursery in Saskatchewan. Asterisks denote significant (**p < 0.05) improvement over non-treated control, based on Fisher's Protected LSD.

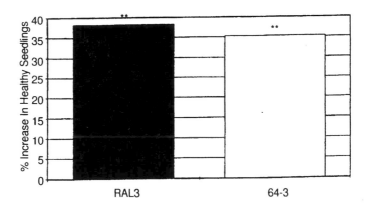

Plant Growth Promotion

Bacterial strains RAL3 and 64-3 significantly increased survival of white spruce bare-root seedlings planted on a reforestation site in Saskatchewan by 19 to 23% when compared to non-treated seed-

FIGURE 3. Percent survival of bareroot white spruce seedlings treated with strains RAL3 and 64-3 before planting on a field site near Duck Lake, Saskatchewan. Asterisks denote significant (*p < 0.10, **p < 0.05) increases in survival compared to non-treated control, based on Fisher's Protected LSD.

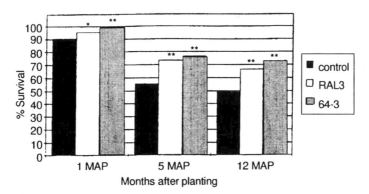

FIGURE 4. Effect of bacterium inoculation on new root growth of Engelmann spruce seedlings on a field site near Prince George, B.C. Asterisks denote significant (*p < 0.10, **p < 0.05) increases in new root growth compared to non-treated control, based on Fisher's Protected LSD.

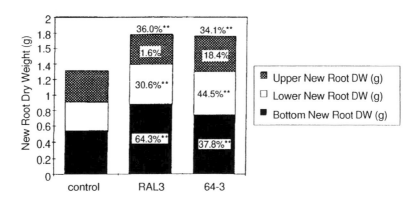

lings 12 months after planting (Figure 3). These strains also in-creased new root growth of Engelmann spruce seedlings planted in British Columbia (Figure 4). Compared to the non-treated control, strain 64-3 increased the dry weights of bottom, lower and upper new roots of white spruce seedlings by 37.8, 44.5, and 18.4%, respec-

FIGURE 5. Increased total plant dry weight of white spruce seedlings inoculated with strains RAL3 and 64-3 on various field sites. Asterisks indicate significant (*p < 0.10, **p < 0.05) increases compared to non-treated control seedlings, based on Fisher's Protected LSD. Prince George and Tappen were assessed four months after planting; Waskesiu and Carrot River were assessed 15 months after planting.

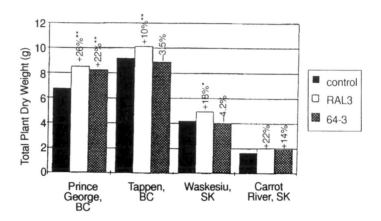

tively. Strain RAL3 increased the dry weight of bottom new roots by 64.3% and lower new roots by 30.6%. Total dry weight (biomass) of white spruce seedlings was increased by RAL3 at three of four field sites, and by 64-3 at one of four field sites (Figure 5). Furthermore, strain RAL3 increased seedling height at three of the sites and stem diameter at one of the sites (data not presented). Strain 64-3 increased height and stem diameter of the seedlings at only the Prince George site.

Shelf-Life of Bacteria in Formulation and on Conifer Seed

Strains RAL3 and 64-3 maintained high populations when stored in commercial liquid formulation at 5°C. The two strains had initial populations of log 9.3 to 9.9 cfu/mL (Figure 6). After 12 months storage at 5°C, RAL3 maintained a population of log 8.6 cfu/mL, while 64-3 remained viable at a population of log 8.8 cfu/mL. At room temperature, populations for both strains dropped below log 8 cfu/mL after 6 months storage, and below log 7 cfu/mL after 12

FIGURE 6. Survival of strains RAL3 and 64-3 in commercial liquid formulation after prolonged storage at 5°C or 22°C.

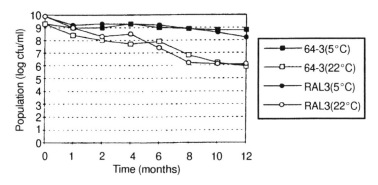

FIGURE 7. Survival of strain RAL3rif on various conifer seeds stored at 5°C.

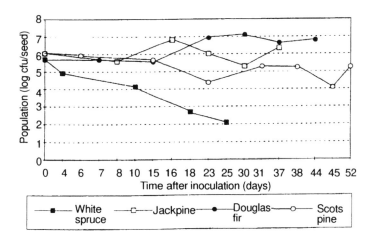

months storage. Strain RAL3rif survived well on conifer seeds stored at 5°C, although some varietal differences were evident (Figure 7). Bacterial populations were maintained at about log 5 cfu/seed for the entire sampling period, except on seed of white spruce where the population declined to log 2 cfu/seed after 25 days. These results indicate that in commercial formulation, both strains maintain high populations at 5°C or 20°C for up to 6 months, but to extend the shelf-life beyond 6 months the strains should be stored at 5°C. Furthermore, seed treated with bacteria can be stored at 5°C for a short period before seeding in the nursery.

DISCUSSION

Two out of approximately 500 bacterial strains, RAL3 and 64-3, have been selected for the ability to suppress Fusarium and Pythium diseases and promote seedling growth. Inoculation of the strains onto Douglas fir seed reduced the incidence of disease caused by these two common fungal pathogens and increased the number of healthy seedlings in a commercial nursery. In tests at replant sites, root-dip inoculation with the strains increased new root dry weight, total plant biomass, and survivability of transplanted seedlings. Seedlings with more roots generally have increased incremental height and diameter growth and it is these seedlings that establish most successfully after transplant. Barnett (1980) working with conifers found collar diameter and shoot dry biomass of container grown seedlings to be significantly correlated with field survival and growth.

In natural environments, growth and yield of plants depends on the quantity and balance of water, minor nutrients, air, light, and heat, but are also subject to positive and negative influences of various rhizosphere microorganisms. Both direct and indirect mechanisms have been suggested to explain the positive influence of certain bacteria on plant growth. Hypothesized direct mechanisms are that bacteria elaborate substances that stimulate plant growth, such as nitrogen, plant growth hormones and compounds that promote the availability of phosphates in the root zone. A popular hypothesis for an indirect mechanism is that populations of various pathogenic and deleterious microorganisms that affect the root system are reduced by the introduction of a beneficial organism via seed or root inoculation (Reddy and Rahe 1989). Each of these hypotheses suffers from insufficient supportive data. Direct information about the activities and interactions of microorganisms in natural soil and plant root environments is technically difficult to obtain due to the complexity and variability of these environments.

Regardless of the mechanisms of biological control or growth promotion, our results have implications for management within the forest industry. Seed inoculation with bacteria capable of stimulating emergence would have obvious benefits in reducing costs associated with poor seedling emergence in commercial nurseries. The inoculants may also be useful for the production of seedlings with higher root to shoot ratios. Our results are consistent with other studies that have shown new root growth to be extremely important in the establishment of outplanted conifer seedlings (Grossnickle and Blake 1987).

There are many opportunities for the application of microbial inoculants in forestry, but gaps remain in our knowledge of how factors such as soil type, soil moisture, soil pH, and silvicultural techniques affect interactions between microbial inoculants and plant roots. There is also a great deal to be learned about the interaction of microbial inoculants with mycorrhizae and other soil biota. As we learn more about the ecology of these microflora, we may be able to establish the critical processes and specific roles performed by different microbes in maintaining sustainable forests.

The results obtained from this and other studies demonstrate that microbial inoculants can be used operationally in container and bareroot nurseries to significantly improve seedling quality. Our reforestation trials have shown that survival and establishment of seedlings can be significantly improved through treatment with bacterial inoculants. The cost of inoculating seed or seedlings with these microbes represents only a minor portion of the total tree planting expense and high seedling quality is an obvious key to successful reforestation. The technology developed through this pioneering project is being expanded to other host species, forest applications and geographic locations. Our goal is to make this technology available to nurserymen, foresters, Christmas tree growers, and other land managers for use in a sustainable forest management system.

REFERENCES

Baker, K. F. and R. J. Cook. 1974. Biological control of plant pathogens. San Francisco: W.H. Freeman and Co. 433 p.

Barnett, J. F. 1980. Density and age affect performance of container loblolly pine seedlings. Southern Forest Exp. Station Research Note SO 256. 5 pp.

Campbell, R. 1989. Biological control of microbial plant pathogens. Cambridge University Press. 218 p.

Grossnickle, S. C. and T. J. Blake. 1987. Water relations and morphological development of bare-root Jack Pine and White Spruce seedlings: seedling establishment on a boreal cut-over site. Forest Ecology and Management 18:299-318.

Reddy, M. S. and J. E. Rahe. 1989. *Bacillus subtilis* B-2 and selected onion rhizobacteria in onion seedling rhizospheres: effects on seedling growth and indigenous rhizosphere microflora Soil Biol. Biochem. 21:379-383.

Reddy, M. S. 1991. Biological control of plant diseases. Pages 33-42 in A. S. McClay eds., Biological Control of Pests in Canada, Alberta, Canada, 136 pp.

Reddy, M. S., S. E. Campbell, S. E. Young and G. Brown. 1991. Role of rinzobacteria in greenhouse mix for the suppression of damping-off of cucumber seedlings caused by *Pythium ultimum*. Can. J. of Plant Pathology 13:284.

Reddy, M. S. and Z. A. Patrick. 1992. Colonization of tobacco roots by a fluorescent Pseudomonad Suppressive to black root rot caused by *Thielaviopsis basicola*. Crop Protection 11: 148-154.

Reddy, M. S., P. E. Axelrood, S. E. Campbell, R. Radley, S. W. Storch and R. C. Peters. 1993. Microbial inoculants and the forestry industry. Canadian Journal of Plant Pathology 15:315.

Reddy, M. S., P. E. Axelrood, R. Radley and R. J. Rennie. 1994. Evaluation of bacterial strains for pathogen suppression and enhancement of survival and growth of conifers seedlings. In: Improving Plant Productivity With Rhizosphere Bacteria (Proceedings of the Third International Workshop on Plant Growth-Promoting Rhizobacteria). Edited by M. H. Ryder, P. M. Stephens and G. D. Bowen. pp. 75-76. CSIRO Division of Soils, Adelaide, Australia. 288 pp.

Reddy, M. S., R. K. Hynes and G. Lazarovits. 1994. Relationship between in vitro growth inhibition of pathogens and suppression of pre-emergence damping-off and post-emergence root rot of white bean seedlings in the greenhouse by bacteria. Can. J. Microbiology 40: 113-119.

PART FOUR:
PANEL DISCUSSION

Prince Albert
Model Forest
Association Inc.

MODEL FOREST
NETWORK
RÉSEAU DE
FORÊTS MODÈLES

Panel Discussion

JANNA KUMI

This is the last section of what has been for me a most interesting conference. I have thoroughly enjoyed the interactions and discussions with so may of you these past few days. I do want to thank the organizers–not only for putting on such a world class conference–but also for inviting me. I hope the invitation was not only because I am a representative from British Columbia, but also because I am a woman. We have heard a diversity of viewpoints this week about building partnerships with aboriginals, First Nations, communities, governments and others. I have not heard the voice of women. It is often said, that the implementation of sustainable development will be done by women on the ground. Therefore, I urge you when you go back to your positions, your jobs, your countries, that you do not forget the role women must and will play, in achieving sustainable forestry.

I have been asked to speak today on public policy under the

Janna Kumi is Assistant Deputy Minister, Forest Operations, British Columbia Ministry of Forests, British Columbia, Canada.

Peter Etheridge is General Manager, Fundy Model Forest, New Brunswick, Canada.

Hamish Kimmins is Professor, Faculty of Forestry, University of British Columbia, British Columbia, Canada.

Steve Smith is Chair, Woodland Section, Canadian Pulp and Paper Association.

Harvey Locke is President, Canadian Parks and Wilderness Society.

Russell Diabo is Advisor, Algonquins of Barrier Lake, Quebec, Canada.

Lutz Fähser is Director, Stadtforstamt Lübeck, Lübeck, Germany.

[Haworth co-indexing entry note]: "Panel Discussion." Co-published simultaneously in *Journal of Sustainable Forestry* (Food Products Press, an imprint of The Haworth Press, Inc.) Vol. 5, No.1/2, 1997, pp. 309-355; and: *Sustainable Forests: Global Challenges and Local Solutions* (ed: O. Thomas Bouman, and David G. Brand) Food Products Press, an imprint of The Haworth Press, Inc., 1997, pp. 309-355. Single or multiple copies of this article are available for a fee from The Haworth Document Delivery Service [1-800-342-9678, 9:00 a.m. - 5:00 p.m. (EST). E-mail address: getinfo@haworth.com].

broader heading of Sustainable Forest Management in Canada–Are
We Heading in the Right Direction? My answer to the question,
"Are we heading in the right direction?" is an unequivocal "*Yes!*"
In support of my answer, I will give you my perspective from the
Province of British Columbia.

I believe there is an international agreement that sustainable forestry must be built on sound economic as well as environmental
practices and be socially acceptable. How to achieve that balance
between what is socially acceptable, as well as ecologically and
economically sound, is the challenge we all face. In British Columbia we are working towards sustainable forestry by conserving
biological diversity, ensuring long-term productivity by conserving
soil and water resources and by restoring damaged forests. We are
investing resources back into the forest and building partnerships
with industry, First Nations, local communities and the public at
large who participate in developing plans for managing the forest
for the multitude of products it gives us.

Quite frankly, the public sees sustainable forestry as a moral
issue. To meet this moral issue, we have attempted to develop a
comprehensive program of initiatives to promote responsible stewardship of the province's rich environmental heritage, and ensure
the long-term well-being of our resource-based communities.

In brief our policies are directed towards building a world class
forest resource sector, with greater certainty over land and resource
use and in finding a balance between economic development and
environmental protection. We have:

- developed our first-ever Forest Practices Code, which sets new
  standards to ensure that the forest resources are managed on a
  sustainable basis;
- a new Forest Land Reserve Act, which will protect British Columbia's privately-managed forest lands, from urban development;
- an innovative Forest Renewal Plan that takes part of the
  money from fees government charges for the right to cut trees
  and uses that money to improve forest management and increase the social and economic benefits provided by our forests;

- instituted a Protected Area strategy, a process that will result in a doubling of the amount of protected area in the province to 12%;
- embarked on a Timber Supply Review which will bring the allowable annual cuts within the province into line with current forest management practices;
- established strategic, long term, regional land use planning forums which delineate management zones–from protection, to special management, integrated use and intensive management zones.

Action has also been taken on developing a fair and open process to settle land claims with First Nations. Through the British Columbia Treaty Commission, British Columbia is moving towards a firm and equitable settlement of treaties which can create long-term social and economic stability for aboriginal and non-aboriginal peoples alike.

That is the policy as it relates to sustainable forests. Now, how do we implement this?

I don't have time here today to discuss all these policy initiatives in any detail, so I will take but three of them as illustration: land use planning, the Forest Renewal Plan, and the best for last, the Forest Practices Code.

Land Use Planning

Our province has a long history of conflict over land use. The debate over which areas of the province should be available for harvest and which ones should be preserved has long been a heated one. It was obvious that the public, on whose lands the forest industry largely operates, wasn't pleased with what they were seeing. They wanted a voice in determining what was happening on the landscape.

Today, it is not often that a decision is made about how land will be used, that has not gone through some form of public participation. From broad strategic decisions to having a say in what happens on a particular piece of ground, the public makes its voice known.

On the broad, strategic level, the Commission on Resources and Environment (C.O.R.E.) was set up early in 1991 to give the public a key voice in land use planning. We have found, that making land use decisions through public consultation is a lengthy, and difficult process. But solutions have been found to many of the conflicts that have been ranging between opposing groups for decades.

Land and Resource Management Plans

Land and resource management plans are further examples of grass roots participation by members of the public in determining land use decisions. There are 12 in progress throughout the province and the aim is to produce the best possible zoning of forest land leading to certainty for all groups.

As well, through a process we call the *Protected Areas Strategy*, British Columbia will double the parks and wilderness areas from 6 to 12%. The key objectives of the Protected Areas Strategy is to protect viable, representative examples of natural diversity in the province, including rare and endangered species and critical habitats. Today, the total amount of parks protected is about 8.2% or about the size of Bavaria.

Of course, we are proud that the World Wildlife Fund gave British Columbia an A − in leading our nation in an effort to protect wilderness. We will reach our goal of protecting 12% and hopefully earn that A+!

Forest Renewal Plan

This plan is a major, long-term initiative to renew British Columbia's forests, safeguard forest dependent jobs and get more value and jobs from each tree harvested. This renewal plan is a partnership between the forest industry, its workers, environmentalists, First Nations, communities and government in order to reinvest and renew British Columbia's forests and forest industry.

It is anticipated that over the next five years, approximately $2 billion will be invested back into the forests and in forest dependent communities. Funded by increased stumpage that industry pays for trees harvested on public lands, the renewal plan has two major, long-term priorities: Reinvesting into forest management and cleaning up environmental damage caused by past logging and road building practices.

The Forest Practices Code

The Forest Practices Code comes into effect on June 15, 1995. The goal of this Code is to legislate the long-term sustainable use of forests. Its guiding principle is ecosystem protection. It took almost three years, and the efforts of hundreds of people, to make this Code a reality. We consider this Code a living document, it will change over time.

The Code will restrict clearcuts, minimize soil erosion and give better protection to streams and threatened plant and animal species. The Code bans logging from within 20 to 50 metres of fish-bearing streams. Clear cutting is restricted on sensitive sites such as steep, unstable slopes, and is banned on visually sensitive areas.

We have also set up an Enforcement Branch within our Forest Service to ensure that the regulations are obeyed. There are 66 guide-books to aid foresters in prescribing everything from timber management to silviculture and assessing the visual impacts of logging.

To make this Code a reality on the ground will take more than a set of rule books. It will take the dedication of all the Forest Service as well as industry staff. Within the Forest Service we have spent over 13,000 person training days to train over 5,000 staff on what the new regulations say. We haven't even begun to train staff in how they will be implementing the job on the ground. On the industry side, we estimate that over 56,000 workers need to be trained. Certainly, in British Columbia there has never been such a training initiative. But what it says to me, ladies and gentlemen, is that for all of us trying to make sustainable development a reality, will require a dedication to life-long learning.

In conclusion, our plan is to take British Columbia into the next century by focusing on:

- Strict new laws and enforcement under the Forest Practices Code,
- Securing the working forests under the Forest Land Reserve to protect the jobs that communities depend on,
- Putting more dollars and resources back into the forests, and creating more jobs through the renewal plan,
- Protecting biodiversity through the Protected Areas Strategy and the Forest Practices Code,
- Setting realistic, sustainable harvest levels through a Timber Supply Review,
- Creating certainty over resource use through public land-use planning initiatives.

With these initiatives, British Columbia is living up to international commitments made at the 1992 Earth Summit to protect biodiversity, promote sustainable forestry, and involve communities, stakeholders and aboriginal peoples in decision-making. I believe that we in British Columbia are beginning to convert sustainable development from an environmental catch phrase into a reality.

Have we reached sustainable development? No, I don't think so. But I want to pick up on what Matt Heering said on Monday: "We need not hope, in order to undertake; we need not succeed, in order to persevere." If, while driving a car, we only look in the rear view mirror, objects always look closer than they really are. We must look at the hood ornament that points forward. In British Columbia, I am proud to say, we have stopped looking backwards, and are now moving forwards.

PETER ETHERIDGE

Thank you Janna. To represent the academic viewpoint as we are moving in the right direction, we have Doctor Hamish Kimmins from the University of British Columbia. He's a professor of forest ecology and is interested in the sustainability of forest ecosystems, ecological site classification, the impacts of management on ecosystem processes, and the modelling of forest ecosystems. His particular concern is that there has to be a sound, biophysical foundation on which we can develop a sustainable forestry that addresses social and environmental concerns. Dr. Kimmins, please.

HAMISH KIMMINS

Thank you, Peter, I also am very pleased to be here and to have a chance to participate. I have learned a lot from the conference. It has been very stimulating and has given me the chance to present my perspective.

As Professor Lal said on Tuesday afternoon, it is very hard to answer this question because it is multidimensional. There is no simple answer. And, in addition to the multidimensional nature of the question, the answer will be different in different part of Canada. We are ecologically different, culturally different, economically different, and socially different in different parts of our country. There can be no one simple answer, no one simple rule, no one simple solution to the attainment of sustainable forestry, however you define it, that applies equally in the many different parts of Canada.

The same is equally true of any one part of Canada, especially in the most ecologically diverse part, British Columbia, where we have up to 75 percent of some aspects of Canadian ecological diversity. We simply can't have a single approach to sustainable forestry in such an ecologically and biologically diverse region. The great mistake foresters have made in the past was to go into this incredibly beautiful and diverse environment and ignore that ecological diversity by practising one method of managing forests everywhere.

We cannot accept that approach, and we must be very careful as we change forestry practices to achieve sustainable development that we don't repeat that mistake. In replacing the wrong approach of the past, we *must* avoid the use of one single other approach everywhere, which would also ignore the ecological, cultural, social and economic variability within our province and across Canada. So my answer is a qualified "Yes"; qualified because there is an ever-present danger that for economic, political or public pressure reasons we might go off track and repeat the mistakes of the past.

To consider this question of where we're going, let's think about where we have come from, and what has been the path of the change. The first stage of forestry in Canada, as elsewhere in the world, was unregulated exploitation. There's nothing wrong with unregulated exploitation where the numbers of people and their technology are such that they cannot take more than nature can easily replace. Exploitation becomes unacceptable when numbers of people and their technology are able to exceed the resiliency of ecosystems and the ability of ecosystems to replace those values. When this happens, the stage is set for the development of forestry.

The first stage of forestry in response to unacceptable consequences of unregulated exploitation is a set of rules and regulations (administrative forestry) that attempts to conserve and sustain desired values. It is human kind's first attempt at forest conservation.

However, it has usually failed because the initial rules and regulations have generally failed to reflect the spatial and temporal diversity and variability in the ecology of the values that are to be conserved: this stage of forestry generally fails to achieve its conservation objectives.

The next stage is ecologically-based forestry, where the policies, rules and regulations are modified so that they are sensitive to the ecology of the various values that we want to conserve. Ecologically-based forestry will sustain many different biophysical values. But it will not necessarily conserve the values desired by wealthy post-industrial societies: spiritual values and aesthetic values. I recognize that these values are not limited to post-industrial societies, but they certainly are a feature of post-industrial societies.

Because ecologically-based forestry will not necessarily sustain these non-material values, forestry proceeds to the final stage: social forestry. This is not only ecologically-based but also addresses the diversity of social, cultural, and other values that society desires from its environment.

All these stages have occurred in British Columbia and other parts of Canada. Forestry evolved first from the old, exploitive forestry to administrative forestry. This change often failed to alter the negative environmental impacts of forestry, and in some cases may have exacerbated them, because the administrative, non-ecological approach failed to recognize and respect the ecological variability of the landscape. British Columbia and some other parts of Canada then moved to ecologically-based forestry, which starts with the recognition of the spatial biophysical variability of the resource, in the form of ecological site classifications and site-adapted silviculture.

All our attempts to achieve sustainability of cultural, social, and economic values, must be within the framework created by ecological site classification. However, while this classification must be the foundation, it will not guarantee sustainable forestry. Ecological science cannot do everything for you. However, without a sound basis in ecological sciences it is unlikely that you will achieve your sustainability goals.

Within this framework of ecological variability, we must pick the disturbance regime and the management and harvesting regimes that match the ecology of the values that are to be sustained. In contrast to some suggestions by earlier speakers, I believe that our current understanding of ecosystem functions and dynamics supports the assertion that all of the classical silvicultural and harvesting systems are appropriate somewhere within our ecologically diverse forest landscape. All of the traditional systems, and some new approaches, from individual tree selection through to clearcutting, if done appropriately for the site and the desired values, are sustainable and environmentally appropriate within the context of the values to be sustained in that particular ecosystem.

Ecologically-sustainable forestry does not necessarily produce the giant trees and the spiritual values of unmanaged old growth that are desired by many in society. It does not, on its own, ensure that these values are passed on to future generations. We tend to argue a lot about whether this is an issue of clearcutting versus partial harvesting. However, I would suggest that in many of Canadian west coast old growth forests, any forest harvesting will detract from the very special value provided by unmanaged "old growth." The issue is, therefore, more one of land use, than an issue of forest practice. We have to decide how much of this old growth condition we want left aside in its unmanaged condition to pass onto future generations, and not argue about whether the old growth is going to be clearcut or partially harvested. This is an important but separate question.

Substantial areas of old growth forests have been reserved on the west coast of Vancouver Island. However, it is very important that we don't allocate all of the 12 percent of coastal forest that we have decided to reserve, to Vancouver Island. There are many ecologically important ancient forests elsewhere in coastal British Columbia that have no reserves. While the public would like to have lots of wilderness and lots of old growth close to urban centers, as an ecologist I believe that we should allocate the 12 percent over the whole of British Columbia in a way that represents all of the genetic populations and all of the ecosystem types in the province.

We must not allow our land use strategy to be driven only by recreational and aesthetic considerations. We must be influenced by conservation, genetic, and ecological considerations as well.

So where are we in Canada? Well, we are a young country, we

have only come out of unregulated exploitation relatively recently. In much of Canada, we only really got fully into administrative forestry about 40 or 50 years ago. We only started the transition from administrative forestry to ecologically-based forestry within the last 10 or 15 years. So it is not surprising that on the landscape you can still see many examples of unsuccessful and unacceptable administrative approach, and sometimes still the evidence of exploitation.

In Canada, we are just beginning to enter the third stage, but all of a sudden we have a case of Alvin Toffler's "Future Shock." Future shock is when the institutions that serve society fail to change as rapidly as society's expectations of those institutions. That is what is happening in Canadian forestry and forestry around the world. All of a sudden the public wants us to be in the fourth stage and to see all of the landscape reflecting that philosophy. Whereas, in reality, we're just coming out of administrative forestry and just entering the ecological-based stage of forestry.

And here is one of my qualifications on "Yes" as the answer to the question. There is a danger that political, economic and inappropriate public pressure could cause us to deviate from what I, as an ecologist, believe to be the appropriate evolution of forestry.

So what are some of the key issues that we must think about? We must put Canadian forestry into an international context. We are stewards of 10 percent of the world's forests, we provide more than 50 percent of the world's international trade of softwoods. We have an international responsibility with respect to atmospheric chemistry, carbon storage, biodiversity, and the provision of wood products to the world market. We must also think of the role of not only all ethnic groups, but also both genders in the development of sustainable strategies. We must, in those forests which are naturally disturbance driven, respect the ecological role of disturbances that, for social reasons, we no longer accept or permit.

The evidence is outside the window. Forest fire has determined the character, the measure of biodiversity, and the productivity of many Canadian forests. We try to put those fires out, sometimes unsuccessfully. Where we are successful, we have an obligation to future generations to try to duplicate through our management practices the ecological impacts of the natural disturbances we have prevented. Otherwise, we will not pass on to future generations forests of the same biodiversity, forests of the same productivity, forests of the same character: they will be different. If we decide to pass on different forests, then it is fine not to respect the ecological role of disturbance. But if our ethical responsibility to inter-generational equity is to pass on similar forests, then we must understand the role of disturbance, and duplicate the effects as closely as we can where we alter natural disturbance regimes.

The following slides show fire and insect outbreaks in the south/southeast of British Columbia, and wind effects in the rain forests of the west coast of Vancouver Island. On the left of the slide, the dense stand is a natural, unmanaged stand derived from windthrow; on the right is the old growth forest not disturbed by wind. The next slide shows a huge natural landslide in a windy, very wet mountainous area on Vancouver Island. These disturbances are what have created the framework for evolution and biological diversity in our forests, and if we believe that we should pass on similar forests to future generations, we must sustain the effects of these disturbance regimes even though we may control the actual occurrence of natural disturbance.

Another issue: foresters must become more knowledgeable about the ecology of the values that society requires them to sustain. Throughout Canada, there is a lack of ecological understanding; understanding of soil is particularly inadequate. In many forests there is an urgent need to upgrade the knowledge by foresters of the ecology of our forests, the ecology of natural disturbance, and the ecology of the values desired by society. Foresters must understand the need to sustain particular wildlife habitat values.

We have to understand that sometimes we need patches of standing trees and snags retained within clearcuts. There may be a need for patches to be left to provide future supplies of coarse woody debris and certain types of wildlife habitat. We have to understand the need for snags for certain wildlife species and ensure a future supply when this is needed.

We must also understand that whole-tree harvesting, a very common practice that in many cases has been driven by public pressure against what has been perceived to be wasteful and untidy forestry, is an inappropriate practice in many of our forests. The needles, the branches and the upper parts of the tree stem should be left in the forest to feed the soil and to maintain site productivity. Whole tree harvesting has often been driven by inappropriate public response to what they have perceived to be ugly and environmentally wrong. But that was often a wrong conclusion and led to wrong government policy. This is one example of my concern about how easily our path forward towards sustainable development can get sidetracked because of inappropriate policy.

Tenure systems: very, very, important. Good, stable tenure systems will not guarantee sustainable forestry, but, on average, if the forest organization or community that is managing forest land has a long-term, stable tenure and appropriate economic environment, it will practice better stewardship than under alternative tenure situations. So, appropriate tenure systems, whether it is community forestry or industrial forestry, are extremely important. We need appropriate and improved tenure systems.

And we need to look at the landscape; much of the focus of forestry—as has been said several times—has traditionally been at the stand-level. I am more of a stand-level ecologist than a landscape-level ecologist, but I completely accept the importance, the overriding importance, of the landscape scale: culturally, socially, economically, and environmentally.

So, in conclusion, we have an enormous responsibility in Canada as stewards of ten percent of the world's forests, and 14 percent of the softwoods. We will only fulfill that obligation if we do respect forests.

We have heard that several times, but I have a slightly different interpretation than some people, as to what "respect" means. Respect means observing and "listening" to nature and understanding the role of disturbance and the ecology of the values that are involved. It does not always mean soft footprints in the woods. It does not always mean small-scale disturbance, if the ecology of that particular value or that particular ecosystem is one that requires disturbance. So, we must respect forests, we must respect their processes, but respect and passing onto future generations these kinds of values does not necessarily mean beautiful to us, comfortable to us. That, I suggest, would be an indulgence on the part of our generation for many of Canada's forest ecosystems. Thank you.

PETER ETHERIDGE

Thank you Hammish. The next view to be represented is one from industry, and will be done by Steve Smith. He is the Chairman of the Woodlands section of the Canadian Pulp and Paper Association and views to be represented here today will be from that organizations. His current position is Vice President/Manager of Timberlands and Weyerhaeuser Canada that some of us experienced by visiting yesterday some of their land holdings in Prince Albert. He's been in Saskatchewan for 25 years, active in forest management, industry, and in government in the provine. Steve, please.

STEVE SMITH

Thank you, Peter. Ladies and gentlemen and visitors to Saskatchewan, welcome. I hope many of you did take the field trips that looked at the Forest Habitat Project and Clarine Lake Demonstration Forest which, for example, is part of Weyerhaeuser Canada's operations.

We think we have some good things to show, some good things to talk about. But you also, those of you who were on those trips were witness to, and *are* witness to, one of the large natural elements that affect forests in this part of the country.

And just as a way of update, on Weyerhaeuser's operations here in Saskatchewan as of today, in the last three days, we have lost somewhere between 50 and 75 thousand hectares of commercial timber. We have ceased all of our forest operations and virtually anybody in our company who has any expertise whatsoever in forest fire fighting, is working on that fire. Which also explains why I'm here.

The Canadian Pulp and Paper Association, for our visitors, is the single largest industry association representing the pulp and paper sector in Canada. It represents approximately 60 companies who produce over 90 percent of forest products in Canada—a volume of forest products estimated in the value of some 40 billion dollars in 1994, about 3/4 of which is exported. Obviously, as a result of that, Canada plays a very, very significant role in supplying global forest products markets.

In response to the question of "Are we headed in the right direction?" I would say, quite definitely "Yes," and I'll come back to the qualifying question in a couple of minutes. A very qualifying question. As an example of some of the achievements, some of the measures we are achieving in the sustainability of forests, I'd like to touch on a couple of cross section examples. From a manufacturing standpoint in the last 5 or 6 years, the Canadian industry has put a tremendous amount of effort and obviously a tremendous amount of investment into improvements on air and water effluent emissions, particularly on the water side. Many of these have resulted in significant improvements in fibre recovery, therefore extending the recovery and usefulness and yield from our valuable forests. Secondly, recycling capability in this country has doubled in the last five years. Throughout North America there are many more recycling dying plants under construction and a new line of work facilities so there has been a tremendous move toward recycled fibre throughout the industry. Another measure: water consumption measured as an effectiveness of productivity has dropped significantly in the pulp and paper industry.

And fourth, I'll give you manufacturing: also significant increases in the use of biomass wood waste energy or biomass fuels for energy. And the reduction of corresponding fossil fuels.

From the forestry standpoint, the shift and I'd say we're in the midst of the shift, is from the concept of sustained yield, a timber management term, towards sustainability of ecosystem in the similar fashion to that described by Herb Hammond this morning. Some measures of that: many companies are hiring wildlife biologists and ecologists, as part of their planning group, expanding beyond the simple timber management capabilities. Those of you that did witness the Saskatchewan Forest Habitat Project are seeing some of the examples, some of the outcome, of more integrated approach into values other than timber itself.

The Model Forest Program across Canada has been a significant achievement and a step towards balancing the environmental, economic, and social requirements of our industry, and it's not insignificant that many, in fact I think I'm correct in saying, in most of the model forests across Canada, industry and First Nations are partners in those model forests.

I learned a new term this morning "wild harvested products." I'd been calling them non-timber products before, up until this morning. Our company has been involved in a wild harvested products inventory. We're not in the business of wild harvested products, but we have developed relationships between where these things occur, mushrooms, berries, etc., and the timber that occurs in our harvesting plans. This information we've made available to those who are in business of wild harvested products.

In addition, standards for biodiversity conservation: other criteria and indicators of sustainability are being developed at both federal and provincial levels. Industry is a participant and expects to continue. Industry is also formalizing environmental management systems to ensure the consistency and continuity of forest practices occurs across all of its operations, making sure that no matter who is carrying out the forest operation, whether it is across the region, or across the country, that consistency is being achieved.

To measure this consistency in an environmental management system, the industry is promoting a certification system to be undertaken by an independent authority and the Canadian Standards Association is working towards an international certification method by the international organization of standards, I.S.O., under the I.S.O. 14,000 umbrella.

The unasked question I referred to earlier is, "are we moving fast enough?" And as a forest manager, sometimes I think things are moving too fast, that sometimes we are over-reacting to the social element that Peter talked about, and that our decisions lack science and fact.

At other times I feel reasonably comfortable with our pace of movement. I have employees, people who work for me who are more knowledgeable in other elements of ecosystems and biology, who would think that I'm moving too slow, that our company is moving too slow and we should accelerate our pace. At the same time that I encounter this in my own company, I suspect and am sure that in our industry there are companies that are out ahead of other companies and even regions of companies that are responding to the local situations, moving faster in one place than they would be in another.

To really answer the question, "are we moving fast enough,"

I think it will not be people like myself, but the consumers, the marketplace and society as a whole, who will be the judge.

A couple of solutions, a couple of ideas on how we can build and improve on achieving sustainability: I mentioned the Model Forest Program. There have been many things learned, even in the first few years of the Model Forest Program, how people can work together, how information bases can be assembled and employed in forest harvesting and natural resource planning. I think those learnings can be expanded much beyond the boundaries of the model forest.

And secondly, the certification of forest products; I think the sooner we conclude the development of an environmental management system, the sooner we truly will have an independent report card that all of industry will have to stand up to and show its colours. Thank you.

PETER ETHERIDGE

Thank you, Steve. Our next speaker is Harvey Locke, who will be representing the Canadian Parks and Wilderness Society. Harvey is a lawyer, in Calgary, Alberta, practices environmental law as well as been involved in oil and gas litigation there. He is also the national President of the Canadian Parks and Wilderness Society, a non-government organization. He's particularly interested in wilderness protection and landscape in ecosystem integrity. Harvey, please.

HARVEY LOCKE

Most people concerned about forest issues in Canada will have heard of Clayoquot Sound, British Columbia and the logging of temperate old growth forest there. World famous are the ugly, eroded, giant clear cuts on Vancouver Island. Less well known is the amount of logging of boreal forest and other forests on private land in Alberta. In my view British Columbia has responded to forest management concerns in a legislative manner that is far superior to the Alberta response. This has been underlined by the recent surge of logging trucks hauling wood from Alberta forests to British Columbia saw mills.

British Columbia recently enacted the Forest Practices Code. While there is no doubt it could be improved, it is a step in the right direction. It recognizes a forest as an ecosystem which has many values and that a forest is more than just a source of wood fibre. In Alberta, the Forest Act does little to recognize other values. Where the British Columbia Forest Practices Code speaks of forests in terms of wilderness, wildlife, scenery and spiritual and recreational uses as well as timber production, the Alberta Forest Act speaks only of timber, perpetual sustained yield, and the possibility of creating forest recreation areas where activities may be prohibited.

The British Columbia Forest Practices Code also contains provision for the establishment of wilderness areas and for special management areas. This allows for conservation biology principles to be implemented through a network of protected core areas, surrounded by carefully managed buffer zones connected by wildlife movement corridors. This is well exemplified in the 1994 Commission on Resources and Environment report for the East Kootenay area of British Columbia which talks of creating protected areas, and safeguarding important adjoining areas of habitat and wildlife movement corridors through special management zones under the Forest Practices Code. A patchwork of Alberta legislation could be used to accomplish something similar. However, where British Columbia created the Commission on Resources and Environment to design an integrated provincial land use strategy, Alberta has recently moved away from integrated regional planning.

Recent reductions in the allowable cut in British Columbia have created a demand for wood from Alberta to feed British Columbia saw mills. This demand has been met by private land woodlots and wood from Indian reserves in Alberta. With the exception of the Municipal District of Pincher Creek, the Municipal District of Bighorn and the Municipality of Crowsnest Pass, there are no permits required or harvesting rules applicable to logging private land in Alberta. This is clearly because the government perceives the rights of the private land owner to be paramount. British Columbia has not yet acted to regulate logging on private land either.

In a series of extraordinary public statements, Alberta's Ty Lund, Minister of the euphemistically named Department of Environmental Protection, has said that Alberta would not regulate logging the way British Columbia has because British Columbia is a "socialist dictatorship." In a March, 1995 press release he recently announced that he has written to British Columbia's forest minister expressing his concern about their regulations. He also made the remarkable comment that his department has no jurisdiction to regulate logging on private land. This is technically accurate from a bureaucratic organization point of view but very misleading. Through Section 92 of the Constitution Act, provinces clearly have jurisdiction over timber. Whether or not his Department of Environmental Protection has authority because of the way the Alberta Government has chosen to organize itself is irrelevant. To make matters worse, when Mr. Lund said that he had no authority to regulate logging on private land, there was a political opportunity for the Alberta Government to do so because the Planning Act, was being changed at the time.

Another striking example of provincial differences is in protected areas. In March, 1995, Mr. Lund announced Alberta's Special Places 2000 policy which was supposed to be a commitment to a system of protected areas in the province. Instead it allows for industrial activity in "conservation oriented sites." Contrast this to the work of government of British Columbia which has moved systematically towards protecting about 12% of British Columbia's land base in parks and other areas from which industrial activity is excluded. When World Wildlife Fund Canada issued its report card on protected areas in Canada in April, 1995, it gave British Columbia an A −, which was the highest grade, and Alberta an F which was the lowest grade.

The answer to the question "How are we doing?" varies dramatically between Alberta and British Columbia. While British Columbia still has some distance to go to demonstrate there will be meaningful change from past horrors, progress on protected areas has been made and the new Forest Practices Code is in place for the beginning of a more hopeful era in the forest. Action is still required on private land in British Columbia. In Alberta, we are rooted in an ecologically ignorant 19th century view of property rights in which the forest is solely a source of wood fibre rather than an ecosystem. In my view there is a land management crisis in Alberta because of the exploitive approach of the Alberta government towards nature. In British Columbia, there is room for guarded optimism.

PETER ETHERIDGE

Thank you, Harvey. Our next view is from the First Nations group represented by Russell Diabo. Russell is a member of the Mohawk Nation of Khanawaki, Quebec. For the last nine years he has worked as a special advisor to the Algonquins of Barrier Lake, a First Nations group in western Quebec, that has just successfully negotiated an agreement with the Government of Canada and Quebec to develop an integrated resource management plan for forest and wildlife in that area. Thank you, Russell.

RUSSELL DIABO

Thank you. Good afternoon. I'd like to thank the organizers, too, for inviting me out here.

I've listened with a lot of interest to a lot of the presentations that have been made and I thank them for also inviting me to sit on the panel because with the situation in Barrier Lake our experience has been paralleling a lot of that of the Model Forest Network. And we always joked around that we were the unofficial model forest because we were not part of the network; we'd developed independently for various reasons.

Before answering the question, "Are we headed in the right direction?" I wanted to review some things to put some context to that question. I think, first of all, I'd like to talk about the status of the First Nations, because a lot of that has been raised in presentations here.

First Nations are not merely a stakeholder group. Things have changed in Canada. In 1982 the constitution included section 35 which said the existing aboriginal treaty rights of aboriginal peoples are hereby recognized and affirmed. And you may recall throughout the 1980's, there was a series of constitutional conferences to identify and define what section 35, the content and the scope of aboriginal treaty rights were.

Unfortunately, those conferences ended in failure. And it was because, I think, they were trying to deal with such large issues affecting Indians, Inuit, and Metis and that diversity to conceptualize it was too difficult. That was a big part of it, there were other reasons as well. But be that as it may, the political discussions ended in failure and didn't resolve that issue.

And so we continue with this legal and political uncertainty as to the meaning of that clause in the constitution until the Supreme Court of Canada recently handed down the Sparrow decision, which handed out a hierarchy of rights. It said that basically conservation of the species was the most important concern to take consideration of, and that aboriginal, social, and ceremonial use was second. After that, other uses could be taken into account.

So, in other words, they recommended an allocation system; first being conservation of the species. They also set out a test in which to look at where legislation conflicts with aboriginal rights, whether that legislation was fulfilling a valid objective, and there were other considerations in there, too, about whether that was the preferred approach of the aboriginal person involved.

The point is that now the section 35 is going to be applied on a case by case basis across the country. It's not a one-size-fits-all national approach.

And the different historical and legal facts of each First Nation will have to be taken into account in any arrangements that are made regarding land and resources, self-government and other matters.

So that's one thing. First Nations, therefore, have constitutionally protected rights, which puts them in a different status from other stakeholder groups. Also, some of the other developments globally have been, in the mid-1980s, have seen the Bruntland Report, the United Nations World Commission on Environmental Development come down with what we considered, anyway from the Barrier Lake team's approach, some very significant recommendations. One is the concept of sustainable development which they put generally as meeting the needs of the present generation without compromising the ability of future generations to meet their needs. A nice general motherhood statement. They also recommended national conservation strategies, and they also recommended that indigenous people should be given a decisive voice in resource management decisions where these affect them.

These recommendations Barrier Lake took to heart, in fact, pressed the Government of Canada on these recommendations. So we saw those as significant developments globally. I don't think that's been mentioned during the course of this conference. Following that, of course that led to the 1992 Rio Conference where we saw the development of chapter 26 of agenda 21 that talked about indigenous people in communities. There is a lot of work to be done on that, I think within the United Nations systems including indigenous peoples in decision making, including this recent intergovernmental panel on forests which has been established in April of this year.

The other United Nations development is that there is a draft declaration on indigenous peoples' rights which is before the United Nations system for consideration. This is a very significant development because it also affects First Nations in Canada in terms of, if this document does make it to the General Assembly level and become ratified, we will have a document with which to measure various states, activities, and treatment of indigenous peoples, including here in Canada. That will be a very significant development and something which is very much needed, not only for Canada but for other countries, as we've heard from and about indigenous peoples' situations in other countries.

The issue of certification, this has come up time and again in a number of presentations here. I wanted to comment on that because I noticed that in a lot of the presentations the issues of trade and consumption weren't dealt with. That is a global context. More and more, consumers, thanks to a lot of work of a lot of groups from universities and environmental organizations, are raising the alarm in drawing the linkage to trade and consumption. It's our view that consumption is one of the biggest problems and the biggest pressures on forests, boreal forests, all forests of the world.

And consumption of forest products is something that really has to be looked at and somehow reduced. I know the industry may not like to hear that, but the fact is those are the types of pressures we have to take a look at because those are the demands that are affecting the supply side. So, when you talk about supplying timber to saw mills and pulp mills, you are talking about the impacts on First Nations people who are at the front end of that process.

As a Mohawk person, I mean we have seen that since the Europeans first came on the shores, how as the one chart showed, we've gone through the deforestation process. You know, we weren't always iron workers building cities. We did live in a hardwood environment in the northeastern North America. So again, we've been on the front end of that process of industrialization. And we've had to adapt to survive. It's something that we're pretty familiar with, we know about the post-industrial system very clearly. And we know that there's problems. I mean the Bruntland Report laid that out that the human species is in a very precarious situation because they are taking over the natural processes, the biological processes of the planet in terms of their spreading out and impacting.

Just to go to the national a bit: all of the provinces have entered into long-term tenure arrangements now. This has serious implications for First Nations peoples. Although we've had the Indian Act applied since 1876 where there have been attempts to restrict Indian people to tracts of lands, called reserves, Indian people have always continued to use their traditional territories. So their aboriginal treaty rights which are now protected by the constitution, they are now finding themselves in the land use conflicts, and resource management conflicts with forestry companies because of these provincial forestry regimes which have been established largely without consulting First Nations. And you'll see increasing pressure from First Nations to reform those systems because it's affecting their basic subsistence, their ability to put food on the table. This affects a lot of northern aboriginal communities, and it affects Barrier Lake. This is why they led a very strong movement to get changes in their area.

This conflict of aboriginal treaty rights has been, within the province of Saskatchewan, we've seen it within certain First Nations. We know that some First Nations are trying to joint venture with companies and work off reserve and get access to off reserve forest resources and they are running into problems with their community members who use those same lands for traditional activities. The way we see it, it is because the provincial forest acts, the long-term tenure arrangements under those acts, the regulations that go with them, do not provide for First Nations land use. The flexibility is not there. It certainly wasn't in the case of Barrier Lake and I've travelled across the country and I've seen the problem in many other First Nations territories. The provinces are, it seems to be in many cases, the biggest obstacle to looking at reform in those areas, not industry in a lot of cases. I shouldn't say that the industry is completely innocent in this, but there are companies making efforts. Certainly the ones operating in Barrier Lake are cooperating.

Some of the other issues are the cutting practices, silvicultural practices, chemical spraying. Those affect traditional activities of indigenous peoples. Conversion of the natural forest into plantations, again which is linked to the chemical spraying to kill off competing growth. Road access into previously inaccessible areas, the provincial ministries who then regulate fish and game of non-Indians going into those areas, putting even further competition on First Nations for fish and game. These are all related issues to the forestry activities, the forestry infrastructure, which has established the damage caused by big machinery. All of these things are common problems affecting First Nations that need to be addressed. I would say in this context then that the Model Forest Network is a step in the right direction, but it is in a vulnerable position. And I say that because in order for Canada to maintain its leadership globally, this idea of sustainable forest management, its concept has to be defined. And I've attended a number of international forums on criteria and indicators processes and the top-down approach is very problematic from my point of view because there hasn't been a lot of indigenous participation; there's been some but it's been peripheral.

There's a lot of other definitions that are coming in there. In fact, in Canada, the Canadian Council of Forest Ministers just recently refused to accept a criteria number seven to recognize aboriginal treaty rights, which is alarming from my point of view, because given the constitution of protection there, there should be an acknowledgement of that and an effort to try to work it out on the ground because there are outstanding land claims to resolve, and there are definitely rights and interests that First Nations have, which co-management arrangements and resource revenue arrangements can work out to accommodate. And again it's alarming to see reports in the Globe and Mail for example, what's going on in British Columbia, where it's reported that only five percent of the province is going to be up for negotiations, which would be excluding any areas which are covered by tree farm licenses.

This certainly I think is going to be a very difficult management problem to say the least for the Government of British Columbia. If that is really and truly their position.

Just to say a little about the experience that we've gone through because we've done the blockades, we've done the protests. We started out in conflict. We basically went through that from 1988 to the spring of 1993. We got the agreement signed in August of 1991. It took us two years to get the forest minister of Quebec to respect that. But finally in the spring of 1993 we resolved those problems, and we were able to get on with the agreement.

Basically, the agreement is in three phases. Phase one was to collect and compile data, phase two is a draft and management plan, phase three is to negotiate implementation of that plan. We are now in phase two. We have moved to a state of cooperation with the forestry companies and with the governments. We are now able to get a working relationship, we are in a full partnership. In terms of what I was saying about the rights of First Nations not merely being stakeholders, Barrier Lake has a trilateral agreement between the Algonquins of Barrier Lake, Canada, and Quebec. We set up a side table with the forestry companies to consult with them and we have an interim projection arrangement which includes Algonquin monitors that makes sure that the annual cutting plans which are submitted to the community for modification are respected during the forestry operations in the field. From this interim experience that we're going through with these– what we call identifying sensitive zones–and coming up with measures to harmonize, we are able to build that into the drafting of an integrated resource management plan over a 10,000 square kilometre area.

So we are dealing with a large territory, we have 18 C.A.T. holders within that territory, those are forest management agreements and we also have outfitters in that area. And the area is also used a lot by Québécois from the Ottawa-Hull area, and from the Montreal area, as well as Americans. So there are a lot of variables in our situation.

And I'm not just putting it out to the Model Forest Network that they are in a vulnerable position. After all the Model Forest Network is an experiment. Just as we're facing experiment in the Barrier Lake situation. We're operating in the same milieu. We have to come down with a management plan which balances the economic aspects, the concerns of the forestry companies, and other users in the area, as well as protecting the biodiversity of the area, the forest characteristics that the Algonquins need to continue their traditional way of life, because they are looking for cultural diversity, it's not just biological diversity. And it's not simply a matter of putting monetary values to this, and this is what we're grappling with. A lot of the issues that have been presented here we are dealing within that territory.

And we're also in a transition zone because we are in the Great Lakes/St. Lawrence forest area and the boreal forest area. So we have a very complex environment and a lot of variables to deal with.

So just like the Model Forest Network, we're having to experiment to try to come up with a plan which is socially acceptable, environmentally acceptable, and economically acceptable. But it's based on the recognition that the Algonquins are the only permanent community in that area, and they have rights and interests in that area, which other user groups do not have.

So far we have been able to manage it. The agreement has expired as of May 26, 1995, and both Canada and Quebec have already informed us that they will extend the agreement to December 1996. We expect a Cabinet decision from Quebec on that shortly. We don't foresee any problems despite some of the political dynamics between Quebec and Canada.

Just on that final point, interms of the rights of the Algonquins. Internationally, they have already set out that regarding the Quebec separation issue, they have the right to exercise their self-determination as well. Consistent with what the Crees and other First Nations in Quebec have said they can either go with Quebec, stay with Canada, or look at independence as well, in the event of

Quebec separation. It remains to be seen what the pre-referendum and post-referendum situation will be, but Barrier Lake's approach is to keep working in a practical way on the ground and a wait the outcome of that process before making a final decision. Whatever that will be, that will be done through a process that they will set up by themselves. With that I think I'll leave it, and thank you very much.

PETER ETHERIDGE

Thank you, Russell. We've pretty well covered the country it seems, and our last speaker is Lutz Fähser from Germany, and he will give us an international view of conditions in his own country as well as regards to how Canadian forestry is thought of in Germany. Lutz is district forester in the city of Lubeck, Germany, and has special interest in developing forest management systems that are nature based or follow natural processes. They've received some recognition from Greenpeace in this area, and actually Greenpeace is looking at using some of their initiatives as criteria. He will also have comments for us on what the view of Canadian forestry is in Europe, and he'll present his views on model forest in Canada. He's certainly interested in a network to enhance the sustainibility in forests on a worldwide basis. Would you please welcome Lutz.

LUTZ FÄHSER

Thank you, Mr. Chairman, ladies and gentlemen. I'll try to introduce myself a little further, because I think this could be some how symptomatic and representative to the situation of the forestry in central Europe and in Germany itself.

I am responsible, as was mentioned, for a forest district in northern Germany. A very tiny district of 4,500 hectares only. It's like a garden in comparison with the Canadian situation. This district is owned by the city community of 220,000 inhabitants.

My professional curriculum vitae is, as I will try to show, in a way characteristic for the forest sector altogether. It started in the

1960s with the concept of high input forestry. This included natural-ly, pesticides, clear cut, exotic species, and so on. And later, when I began practical work, I went through very high confusion caused by the failure of all these theories and practices I had learned so prop-erly before.

Now finally, I have turned to a low input forestry, which tends to respect reality. It means to respect nature, and nature processes as far as possible. In 1977 I got my specialization in forest economics. But my daily concern now is to detect and follow the ecological and social limits of forest ecosystems.

All the models we learned in university and everywhere do not adequately show the complexity and vulnerability of forest ecosys-tems. But still now, we in Germany, forest professionals, still seek to optimize these models instead of optimizing the super-complex natural situation itself.

Finally, after this short path of life, I came close to a level of awareness to which others here in Canada, and North America like Herb Hammond, I heard this morning with much pleasure, had come before.

With this background of considerations and experience, we de-veloped in the city of Lübeck, a concept of a nature-oriented forest use with very close participation of the citizens.

The philosophy behind this concept is rather simple. Forests, first, are too complex to be understood by only some officials in charge. Forest use means the use of natural products within the sector of primary production. Here in this sector optimal ecological functioning is the pre-condition for economically optimal results and for the fulfillment of social and cultural demands. These two considerations together lead to a third one. Natural processes in forest ecosystems honestly can neither be predicted nor imitated at the operational and instrumental level. And that exactly is the level of forestry activities. But, we have to manage our forests. Timber is a sound natural product. In order to find a solution for this manage-ment we must follow the permanently on-going natural processes, instead of initiating or mimicking them. Interference is to be re-duced to a minimum.

For example, only extensive thinnings, only selective cuttings, mostly natural regeneration, no pesticides, no fertilizers, and no disturbance of the soil.

Parallel with the managed forests, we in Lübeck set up a representative net of 10 percent reference forests, where no interference is allowed except hunting. These reference forests shall serve, in the long run, as a scale for natural dynamics in former managed forests.

Our concept found high acceptance in the community. It led, not to our surprise, to a considerable irritation in the state forest service; and it led, finally, to appreciation by Greenpeace Germany and Greenpeace International.

In 1994, Greenpeace formulated seven principles of sustainable forest management based on our small concept in Lübeck, which were published also in English and even in Canada. And this year, 1995, Greenpeace started a process together with Friends of the Earth–Germany, of acknowledgement through certification by a special green label.

This was the first part I wanted to report, to show you my background for the discussion. The second I wanted to tell you is how Canada's forestry is seen by environmentalists in Europe.

That discussion arose mainly from the critical discussion on Clayoquot Sound is Greenpeace. They are focusing especially on two aspects. The first is the participation of people, especially of First Nations, and the second is to stop any clearcuts.

Similar campaigns are taking place in other countries with boreal forests, like Russia, and Scandinavia. And it is very interesting to see that Sweden in Scandinavia recognized rather soon the power of ecological and social arguments in the public for the economic market, for the business. And Swedish industry is trying, at the moment, to get away as soon as possible from a bad reputation in an ecological sense.

With the Model Forest Program, Canada is starting very promising attempts towards an honest participation of all people concerned with forest management. With respect to clear cut, however, as the predominant harvesting technique, there is, from a European point of view, not much change in sight which could reassure the environmentalists, especially Greenpeace.

The third item I would like to mention is this international conference here today, on sustainable forests. I am personally very grateful to the Canadian Government for my participation in this conference. I felt the strong efforts of presenters and participants to start into a new and responsible forestry, taking into account the actual knowledge and needs of society, including the knowledge, wisdom and needs of First Nations.

Through this knowledge, Canada has an enormous advantage over European forestry. In Europe, nearly all forests are secondary, man-made forests. Indigenous people who lived in forests, no longer exist. It means Europeans lost the holistic and detailed knowledge about nature and forests completely. We lost it completely and forever and it was replaced only by mechanistic scientific data, which led to artificial vulnerable forests.

Within the Canadian Model Forest Program, the participation of means an essential contribution towards future First Nation source of knowledge sustainable forests. The program should be accompanied by research to become partially understood by us, by our modern instrumental language. Research can help to find out at least the limits, the vulnerability of complex living systems.

The further success of the Model Forest Program will depend too, on the documentation and cooperative interpretation of the effects of the system caused by forest management.

I saw in these past days, several preliminary results from field studies, which pointed clearly that clearcut would worsen the situation dramatically. But I felt at the same time, the helplessness of the scientists in considering how to bring these results up to the decision making levels in our society. This conference and the Model Forest Program is indeed a very exciting event, which is, at the moment, unique in the world, and which really can lead to a new understanding of the role of man in forest ecosystems. With this, I think we are heading into the right direction. Thank you.

PETER ETHERIDGE

Thank you, Lutz. At this time I'd like to open it up to discussion to the floor. If you have a question or comment, could you move to the microphone and identify yourself please.

FLOOR DISCUSSION

Keith McLean

My name is Keith McLean, and I'm with the Canadian Forest Service in Prince George.

A number of the presenters this morning have heightened our awareness of ecologically based management and need for that important management. Mr. Hammond certainly brought that to the forefront this morning. However, I'm also faced with this question of what does Canada do as a producer of wood and live up to perhaps a world obligation to supply fibre?

We saw numerous times the population growth statistics, and they are pretty substantial, they are pretty scary. Now it would also seem to me that through ecologically based forest management, we won't be able to harvest wood on a broad scale as we are right now. But what does Canada do? Do we have a world obligation? Perhaps I could just throw that open to the panel.

Hammish Kimmins

Yes I think we absolutely do have a world obligation. For a start, forests are so extensive. The way they are managed is believed to have a significant effect on global atmospheric carbon budgets. And I think we have to think very carefully about the interaction between forest fires and the atmosphere, and between forest harvesting and the atmosphere. So we have a global obligation there.

We also have an obligation, as do all countries, who are stewards of a significant segment of the world's forests, but in the face of a world population that is increasing at a rate that the annual increase in demand for wood products is equivalent to the annual allowable cut of British Columbia. Faced with that reality, and the fact that the population is expected to more than double to in excess of 11 billion people, in the time that it will take to grow another tree crop, where the forest fires just killed trees up here in central Saskatchewan, we have to consider the effects of the land use decisions and the resource extraction decisions and the export decisions we make here for the biodiversity and the people in other parts of the world.

We are a highly regulated forest country. It is not all as it should be, there is much yet to be done. But we are going in the right direction and we are highly regulated, and there is a danger of becoming so highly regulated we might go back to administrative forestry. There is a real danger of that.

But I think we're going in the right direction. If we chose to significantly reduce our exports to the world, then the world market, it seems, would be satisfied. That seems to be the way the world works.

There are lots of countries around the world who would love to cut down their forests and sell them to create capital. And we know that because of the low harvestable volume in the tropical forests compared to the high volume and the high productivity of many of our northern forests, that if we would forego that export economy, we would directly impact the sustainability of many values in the tropical forest. So I think we do have many dimensions to the international responsibility and while we must think locally and seek local solutions, we must consider global locations.

Harvey Locke

I have a completely different point of view.

There are two things I think are really important and I've heard raised a number of times by professional foresters here. The concern that it is really society that is driving our behaviour, and that really it is not fair to place the burden on professional foresters to make these ecologically sustainable decisions given the demand and the market and that sort of thing.

With respect, I think that's a big copout. The public looks to professional foresters and our bureaucratic apparatus and our social structure. That is to say, the people who know best are supposed to be managing. The problem has been that there has become a social mistrust of professional foresters, because it is patently obvious to people when they drive past streams that are full of soil, or massive clearcuts, that simply hasn't happened. And I really think that we have to begin with the responsibility issue at home, not to blame it on some imaginary market that's out there.

And I really disagree with Hammish's argument "let's continue to export lots of wood here to keep up the unsustainable world demand of timber, because somehow we're going to offset tropical forest destruction." Surely the answer is let's get our act together on a global scale, including in Canada. Not let's continue to cut down Canada to try to abate the pace of unsustainable forestry that will happen anyway unless we change that behaviour elsewhere. We really diametrically disagree on that.

Steve Smith

I guess I'm going to join Hammish on his side. I think we do have a responsibility. Fifty percent of the annual wood consumed in the world is consumed for fuel wood, charcoal, home heating fires, and those that occur in parts of the world where 2/3 of the population exists. I think we do have the wherewithal, the science, the responsibility to manage the forests and to meet the rising demands for forest products. At the same time we have the obligation to use our wisdom to do that in a manner that does achieve criteria of acceptable sustainability.

Janna Kumi

I think to place the burden on the professional foresters is to give credit more to this profession than it deserves. I think the answer is in all of us. The professional forester manages the forest for a multitude of reasons and values and must take into consideration economic as well as social implications of decisions within an ecological framework. That does not mean to say that the forester cannot be the forefront of leading the change towards less consumption. But nonetheless, let us not make the forester the culprit. I think it behooves every one of us to take on that burden, and lead the way to a different way of managing our resources and living on this world.

Surely a good part of the solution is reducing the birthrate, and I don't think foresters are enough in this world to make any impact in there whatsoever.

Marjorie Bousfield

I'd like to ask a related question. My name is Marjorie Bousfield and I work with Waskaganish First Nation Crees in northern Quebec. And by the way, Russell, thank you very much, I really enjoyed your presentation. The relatedness, you were just talking about demand, and all these forests being cut, and yet I didn't hear any mention at this conference at all about the use of alternative fibers. So Canadian Pulp and Paper Association, you can make paper of things other than wood fibre. What is going on? How much are you encouraging your companies, and government, and research in universities? What is going on there? Are you really pushing this?

Steve Smith

Right now there is a tremendous amount of work going on with the research facilities in Canada on non-wood fibers as supplements or alternatives. There certainly is or are alternatives to making paper products from things other than wood. They too, also have inherent shortcomings in the area of energy consumption, as well as some of the ones we're familiar with, flax for example and fibre is extremely hard on soil. It is harder on soil than is growing trees. So they are alternatives, they are being examined, they have a role to play. Will they replace wood fibre? I think the answer is "no."

Marjorie Bousfield

Well, what about something like hemp for example, are you pushing the government towards making it more legal to grow the type of hemp with non-hallucinogenic properties? It's not supposed to be hard on soil, it's supposed to grow in all sorts of conditions, and is supposed to be able to be used in many of the current mills with very little change.

Steve Smith

The industry at this point in time is not lobbying government. No. The industry is undertaking research into the use of alternative fibers for the making of paper.

Peter Etheridge

Russell, you have a comment?

Russell Diabo

Yes, I guess this goes back to my comment about the trade and consumption issues which weren't really addressed in this conference. As I indicated from the very right point of view, we see consumption as one of the biggest pressures on the forest, and we think certification is the key to first of all getting a definition of sustainable forest management. We think it has to come from the bottom up, instead of the top down, like we've been seeing in international forum. Once we have some base line data from different forest types in Canada, then I think we're in the position to start looking at broader principles and concepts of criteria indicators for sustainable forest management. But related to that is the issue of educating the public. I know from my trips to Europe, that there definitely is a growing consciousness or awareness of the need to reduce the consumption and waste of forest products, whether that's paper or lumber. And also that along with that consciousness is more and more of an interest in determining where these products are coming from and are they in fact being sustainably managed and what are the impacts on the local people, the indigenous people, and other forest dwellers. And these are the areas we think need more research, and development is looking at ways to get the consumer more aware in buying products which are environmentally sound and, of course, Barrier Lake isn't opposed to logging. We've said that we're trying to work it out. Our management plan is to try to balance environment and economy, but obviously there are going to be limits to what is sustainable, as we move from this concept of sustained yield to sustainable management other factors have to be taken into consideration and those are the reforms that really need to happen. The biggest obstacle that I see to that, potentially, is from the provincial governments and their unwillingness to share decision making with First Nations in areas of resources that concern them. There is some progress being made but there is a long way to go with that. And if we can do that, then I think we can deal with some of these issues about supply and demand. Thank you.

Paul Mitchell Banks

My name is Paul Mitchell Banks and I'm a consultant and PhD candidate with Brooks Young Community Forestry. I've been going to a lot of conferences lately and we're always hearing the same thing that looking back and navel gazing and wondering how we got here and where do we go.

One of the biggest challenges, I think, is trying to transform an industry that's based on high volume, low quality, to perhaps a different volume and different quality.

When I was in banking I was having a lot of my logging contractors coming in and saying that things are changing and I don't know if I'll be able to continue operating. And it's not that there's no forest, it's that I have the equipment that's completely inappropriate for what we're facing. And I thought it was interesting to hear that we have this purported obligation to supply the international market with all the timber we can provide it, when there are other ways of approaching it.

The first is just saying no, that maybe we can clean up our own mess before we start trying to solve the world's problems, but the other is getting a little bit creative.

I spent last summer going through Scandinavia and central Europe looking at small scale forestry and also industrial forestry. You've got countries like Finland, 52 percent of their volume comes from thinning and they've got a 25 million cubic meter increase in their standing volume every year. That's 1/3 of the volume cut every year in B.C. I had mill owners in Germany telling me to send over our antiquated equipment because it was too big for the volume of the pieces that we were handling, because the volume that they had, or the size of the butts were too big now for the equipment that they had.

I think there's a tremendous number of lessons that we can learn but the big thing that I'm beginning to realize is that we can talk, talk, talk, but it's time for someone to say, "I'm not going to do it this way, I'm going to do it differently."

In many ways, we're at the mercy of policy and that's something I'd like to hear more about. In British Columbia, we've got tenure policy that's 40-50 years old drawn up at a time when we thought we'd never run out of old growth. Do you see a way of bringing about tenure reform, and it's not just in British Columbia, it's really across the country, changing the terms, changing the types, changing the owners of tenures. And how do you as the panel propose to address something like that, or is it even appropriate?

Janna Kumi

I think the British Columbia Government has certainly gone on record, and I hope I'm not putting many words in the Minister's mouth when he himself has said that the government expects to get into tenure reform if it receives another mandate. They will probably go for an election later on this year or at the latest next spring.

But regardless, if the government comes back into power, the next government, the Liberals have gone on record saying they will institute tenure reform. So either way, however you wish to vote, you're going to get tenure reform. This can be done in a slow and systematic fashion, or it can be done all cards on the table, reshuffle the deck.

There are tremendous implications though, to communities, to workers, to the industry, and even an NDP government understands the volatility of markets and capital. This is not an excuse to say that we should not embark upon it, but I think it is a statement that whatever we do we must do in a rational and planned manner to ensure that not everybody gets upset and that it can go in an orderly fashion.

Hammish Kimmins

I'd like to address a couple of points there. Firstly I think we have to be very, very careful about adopting the European model in Canada.

I grew up in Europe, I learned European forestry, I love European forests, they are beautiful, they appeal to me because I grew up in them. They are lovely for Europe and God protect Canada from that kind of forestry. We have wild forests, wildlife here eats you for lunch if you're not careful.

The thinning option will give you a forest like the European forests in which there is virtually no coarse debris as we heard from Chris Maser and I agree with the importance of a certain amount of a coarse woody debris in our forests. We know from wildlife and by diversity that snags are extremely important. If we go with the thinning option, we're not going to have those. It's very interesting that the great debate is about clearcutting and partial harvesting and it's always the implication that the partial harvesting is going to be a lot more user friendly. Whether it is or not depends upon the values you want to sustain, the kind of ecosystem you are operating, and the scale and way you do it. But in terms of coarse woody debris and snags, if you have an even-aged forest that comes from clear cutting or shelter wood, and that is run on long rotations with no commercial thinning, you will finish off with more coarse woody debris on the ground and more snags in that forest with better wildlife habitat for a lot of different animals than if you practice uneven-aged forestry with thinning that essentially captures natural mortality. Which is what happens in the European forests and it's what Lutz was saying that they want to get away from. And I agree with him, I think that they are putting back some useful structure into their forests. So we have to be very, very careful about how we adopt that and how we think of the solution.

I'd like to come back to this international thing. And I think Harvey's implying that there is a greater difference here between this than there really is. Nothing I said was inconsistent with recycling, with producing, value added or any of those situations. But I have travelled enough internationally to be very, very concerned about the impact of the world of biological diversity and the world forests of this expanding population. And seeing how the standard of living is increasing and seeing how the increase for wood demand for non-fuel wood is occurring as well, I'm wondering where the wood is going to come from. In many cases, there seems to be two choices: from their own tropical forests which has important economic and cultural advantages for them because it builds up their local economy. So we've got to think about that aspect. But unless that is done in a sustainable fashion and many of those countries, for political, economic, and a variety of other reasons do not seem at the moment able to do that very sustainably, then if we were to seriously restrict our export trade for a variety of reasons, then I suspect, and I've read a lot of other well-informed people who have reached the same conclusion, I suspect that this would have a negative effect on a number of other forests around the world. What I was saying is entirely consistent with many of the things that were said by Harvey, but I continue to insist from my experience of travelling around the world that we have to think very carefully of our international obligations as we think about changing these things.

Peter Etheridge

Russell, you had a comment?

Russell Diabo

Yes, just in relation to changes to tenure systems, this is one of the challenges I see for the Model Forest Network—being able to move the research that's being done within the Network to the practice level, and being able to look at reform into industry practices and government regulations.

Just to give you an example of what we're doing in the Barrier Lake agreement, the scope of our agreement says that when we develop a draft integrated resource management plan and move to phase three which is negotiating the implementation of that plan, we come up with recommendations which may aim at modifying management and exploitation methods, administrative and contractual adjustments, and amendments to regulations or laws. So this is exactly what we're looking at because Barrier Lake was not involved in the development of the Quebec Forest Act or its result in regulations. We're looking at the Barrier Lake Territorial Management Plan which is looking at reforming that whole tenure system. And I think that if the Model Forest Network would take a look at that agreement and that approach and the governments involved in the model forest, particularly the provincial governments and First Nations, that would be a good way to get a government-to-government relationship going for identifying those interests and working them out in some kind of comprehensive way. We think our agreement is a model of sorts to be looked at in that regard for across the country because whether you're talking about the CAPS in Quebec or tree farm licenses or forest management license agreements, they are all very similar in the way that they were laid out. So just to bring that point up that in our process, that's what we're looking at. We have a wide scope. Thank you.

Peter Etheridge

Steve, you have a comment?

Steve Smith

Russell triggered something in my mind when he talked about expanding achievements beyond the model forest.

It certainly was Weyerhaeuser's intent to take the achievements from the model forest and expand them beyond and that is happening now, and while the habitat work was an example of that specifically, where that has now become a part of our harvesting and planning process throughout the area. But coupled with that, the benefit of the model forest was that it allowed government and industry participants to analyze what the effort, what the resources, what the costs were of doing these things in a different way than we had done them in the past in a controlled area before you expand it over a larger expanse of management area. These things do carry obligations and costs with them, and we are going to have to learn how to share those at the same time that we share the benefits.

Lutz Fähser

I think it's not only the problem often for Canada which is here discussed because I think it's really a crisis of forestry worldwide. Because we see all over the world and—more than us—society sees all over the world, that we do not meet the needs of society, and we do not meet the needs of nature.

We are well on the way during the last 200 years towards extinguishing large parts of nature and its processes. So the question is to us what shall remain of this symposium, and are we going to change things, and how are we going to change things. We have normally two solutions, two possibilities to change things quantitatively or qualitatively. And as far as many of the participants saw, we are still at the level of quantitative change. Because we still think that we know how nature is and how nature works and we are only going to fight a more sophisticated anticipation of nature, and we are going to introduce more and different participants in the process of decision making but we are still going to make decisions upon nature though all of us know that nature is not understandable, really.

So that's the quantitative aspect, and we have failed in the quantitative aspect world wide. And I'm rather sure, ten years from now, we will have another slide from Mr. Kimmins where we will have number 5 and we will see the solution did not come from introducing more social input, rather, the solution was qualitative—that is, to respect nature and to try to fit in nature and our basic functioning of nature, otherwise we will ruin it. Not only in the boreal forest but everywhere.

So, it is really the difference you mentioned: Europe cannot be a model for other parts, that is quite true, completely true. We thought that the quantitative aspect, where people think that it is possible to mimic nature by forest activity, by certain instruments, by certain techniques it is theoretically not possible because nature is too complex. We must switch over to the other aspect, to follow nature, to fit into what nature shows us, that means that we have to obey structures of nature.

Here in Canada, it's such a happy situation that there is still a lot of knowledge and you do not have to work with the exotic life that tropical countries do mostly. So my idea is Canada has a good chance to come to a qualitative solution. And within this qualitative solution, for example, forest fires are there and they are there to renew forests, and we can't combat them. Diseases are there, and they are part of nature. Furthermore, these huge interactions are of nature itself and we are not allowed to make additional huge interactions or catastrophes like clear cuts at large scales. Thank you.

Paul Mitchell Banks

Can I just respond to that? Two quick points. One is, I wasn't suggesting European models are all that we should apply here, but I think there's a lot of lessons to be learned. Hammish, you mentioned the lack of coarse woody debris in European forests, that's a past practice. If you walk through the forests right now in southern Germany and throughout Scandinavia, you will see lots of coarse woody debris on the ground, lots of snags. That picture is from Chris Maser's book, *The Redesigned Forest*. It's an out of date picture. So the practices there are evolving to a tremendous degree.

I just think that we have to get a bit more creative, you can't just reduce the size of the clearcut and say it's a new practice. I'm not saying that clearcuts are bad; I'm just saying let's look at the practice in an original sense and we may come back to that decision again, but I don't really think that we're getting outside of the model of the paradigm. We're just slightly changing it.

Hammish Kimmins

Basically, I agree, and you can manipulate either partial harvesting or clear cutting to achieve structural goals. So when I said that traditional European selection harvest has a lot less snags and has a lot less coarse woody debris, my understanding of from what I've seen over there, it's great. That's the traditional system. You can organize partial harvesting to leave that material just as you can organize clear cutting or shelter wood or an even-age system to leave that. So it's a case of deciding what you want, what the values are that you are managing for, what's the ecology of those values and designing a system that will achieve that sustainably.

Peter Etheridge

We have a gentleman at the microphone who has a question. Go ahead.

Floor Speaker [Name Not Audible]

I am from India. This is not a question but [. . .] Look at wood substitution. I just wanted to mention that my industry has taken this step because most ministers coming forward. So we took up the international discussion and some of the decisions were: one, they cannot use wood for construction. There is a specific solution that no wood is used in construction. Some of my friends across the border would not appreciate, but we reduce the customs duty on imports of wood and wood products so that the demand on our products would go down. [*Rest of statement not audible.*]

New Floorspeaker [Name Not Given]

I have a couple of questions, first one on wood certification. I think Russell and Steve, you may be the best people to direct this to. First of all, it's important to note that the Canadian Standards Association was hired by the Canadian forest industry for all intents and purposes to develop a certification effort in response to an international effort called the forest stewardship council which is not affiliated with any interest group, includes industry, includes environmental groups, is a non-profit organization, unlike the C.S.A. Which operates for profit. So it's also important to point out that after participation in the C.S.A. process by a wide range of environmental groups and some indigenous people, that they have withdrawn from that process because they feel that the process is both flawed and a direct conflict of interest. In other words, wood certification or wood certifiers need to be removed from wood producers if they are to have any validity or any competence for the consumer public. My question: would either of you or anyone on the panel, see that problem with the Canadian Standards Association approach? And would you as a representative of the timber industry, Steve, be willing to use the Forest Stewardship Council instead of the C.S.A.?

Steve Smith

Right now the Forest Stewardship Council and the C.S.A. are kind of the two big, but by no means are the only certification schemes that are in use. There are over 100 of them in various places. It's interesting, at the same time that the Forest Stewardship Council is rejecting the C.S.A. approach because it does not accurately represent the environmental interests; that the C.S.A. is claiming that Forest Stewardship Council is not offering seats on the directorship of the bodies with the forest stewardship council. It's a Mexican stand-off, if you will. What I believe is going to happen, is that there is in fact going to be a merging of the two initiatives, I think it's already beginning to happen. I have some notes made in Stockholm last week that suggest that maybe the ice has already been broken, there is a beginning of a convergence towards a common certification scheme as widely accepted as being independent, truly a third party certification scheme.

Russell Diabo

Actually Barrier Lake was one of the ones that was critical of the Canadian Standards Association's approach in developing the criteria for certification.

Our view was that there wasn't adequate representation by aboriginal people. The only organization was the National Aboriginal Forestry Association and Barner Lake is not a member of that organization.

What we're doing with the trilateral agreement, that process has serious implications for us because we think the debate is still on about whether it should be voluntary or required something that should be looked at. As you mentioned, there was a number of certification schemes that are being looked at and being discussed around the world, and we think that certification would be a useful way for the companies that are cooperating with the reforms in the Barrier Lake territory to help encourage them to proceed into that way, and to help in marketing the strengths, for example in the European market, where they are looking at becoming more concerned about the source of these products. That's one way to make it instead of being a liability, an asset–to have certified product that can be marketed in such a way to take a leadership position in the reforms that are taking place. Our view is that we would gladly cooperate with the companies if we can agree on the criteria and indicators in the management practices within the territory. We think that is vital.

We were very concerned that there would be a lot of confusion with the Canadian Standards Association approach. They had an approach where they were trying to get to the International Standards Organization by June of this year and we felt that was much too fast, there was inadequate involvement from other leaders and as Pete said maybe that's what needs to be looked at, a bit more going slower and taking a harder look at the F.S.C. And maybe others. Personally, Barrier Lake has already decided it will take a look at the F.S.C. approach.

They are satisfied with the principles and criteria that have been outlined there. But other aboriginal groups should have that opportunity as well, as well as other user groups. That's our feeling.

Herb Hammond

Steve, I'm not familiar with the meetings taken place in Stockholm, but the Silva Forest Foundation has participated with the Forest Stewardship Council since its inception and is close to bringing forth its first certified wood from the western part of this country, and is also working with other people in North America and other parts of the world on wood certification.

I would suggest that I think it would be quite amazing and I think totally unlikely that you will see a coming-together of the industry with environmentalists over wood certification because, quite frankly, environmentalists are the reasons behind wood certification.

The industry is responding to try to do a short cut upon that particular situation. To have the industry certifying itself, or playing a major role in it, is like the fox guarding the henhouse. And I think that you will find that there is a strong international movement that recognizes that and is not about to be co-opted by slogans and talk.

I think personally, it's in your industry's best interests to stay out of certification because you will end up having a much better role outside of it, showing your goodwill, showing the high quality of practices.

My other question: I heard a lot of discussion about our obligation to keep supplying the amount of wood that we supply. I think we first should recognize where that wood goes. It doesn't go to underprivileged countries. It doesn't go, in large part, to supply houses of a size that people need. It doesn't go to supply paper that people need. As much as anything, it goes to fuel the fires of consumerism, that are not only destroying forests, but are destroying atmosphere and other vital parts of the functioning of this planet.

So to suggest that we are somehow or other helping other people's needs by liquidating our forests and confusing that with the fact that softwood timber is used for entirely different purposes than tropical hardwoods, and has largely distinctly different markets, is confused at best.

I wonder if there is any economic analysis that has gone into these remarks by the panel. I further wonder if the panel knows of cost benefit analysis looking at all values for timber extraction in the boreal forest. For example, not only in Canada, but in Russia that would substantiate this as an economic venture and not a massive subsidization of trans-national corporations.

Janna Kumi

I'm going to let professor Kimmins deal with his topic. I just want to make a point here in terms of . . . I don't know how the rest of the panel shares professor Kimmins' viewpoint.

In terms of me as a panel member, I think I was quite strong in saying that the province of British Columbia has developed a number of initiatives to reduce the annual allowable cuts. Our timber supply review is bringing that cut down quite significantly, and there's no doubt that the Forest Practices Code will have an impact on the timber as well. To negate those effects and to get more value out of every cubic meter that we cut, we have instituted that forest renewal plan. So there's no doubt that timber from British Columbia, the amount that's going to be cut in next decade, is going to be considerably less than what it has been.

Hammish Kimmins

I don't disagree at all that with the point you make about the excessive consumption. We're still bulldozing houses and taking the wood to land fills rather than recycling it or using it for energy. There's a lot of room for recycling and reuse, to reduce consumption.

However, if you look at the three legs of sustainability, every time you make a decision on one leg it is going to have ramifications because it, like an ecosystem, is an interconnected system. There are ramifications for employment, economies, and yes, certainly you can increase the employment per cubic meter of wood cut by going to value added.

I think that is happening now, long overdue, I agree. There is some economic and trade barrier reasons as well as other reasons why things were the way they were. But I still am very impressed, when I look at the age class distribution of the world, when I look at the change from straight communism to a communist form of capitalism in China, I look at the growth and needs of the Indian population. Certainly you're right, construction of houses is not going to be from the tropical hardwoods, it's probably going to be from tropical softwoods grown in plantations in areas that used to carry tropical forests. The fact is, it is one planet Earth, it is the global village, and there are cultural, social, and economic implications of decisions made in one place.

That is the context of my remarks and I believe firmly that there is a global context and I said we must look locally, we must sustain functioning of our ecosystems, we must be sustainable at home, but I don't think we can consider that sustainability as though it were isolated from the rest of the world. It's a world ecosystem socially, culturally, and environmentally. It's all linked, and we have to do what we do in that context.

Herb Hammond

I guess I'm quite familiar with that actually, from my work in Russia, and it's not an economic cost benefit analysis like I was referring to.

There are a few of us that have done cost benefit analysis of boreal forests, that range from not just people with my ilk but the Newfoundland Government, for example, has looked at a lot in Labrador. And when the answer, interestingly enough, I have yet to see a cost benefit analysis that was done with any validity, that showed that timber cutting was other than massively subsidized for boreal forest.

So the question earlier about alternative fibers for paper since much of the wood coming from the boreal forest goes to paper, is a very germane one, and I would suggest to everyone that it's time for society to do basic economic analysis.

I would suggest ecological economic analysis of our activities in the boreal forests because I think we'll find that they don't hold the test of good scientific and economic scrutiny.

And making paper out of alterative products is a readily available alternative. It's only been 200 years since we started making paper in a large scale out of trees. The main reason that we're doing it is because we don't charge what those trees are worth for water, for indigenous culture, for wildlife, for carbon sequestration. So let's not operate an illusion here that there's been a balanced analysis of what happens in the boreal forest.

Hamish Kimmins

I don't know what such an analysis would conclude, Herb. I agree completely that such an analysis should be undertaken for all forests. I have no idea what the results would be.

I would just like to comment on the argument about alternative fibre sources because this often comes up in the context of British Columbia. British Columbia is rather short of agricultural land so there isn't a lot of land currently producing food that could be used for alternative fibers. In much of British Columbia, to grow significant amounts of fibre, you would have to cut forests down.

Herb Hammond

If you lived where I do, Hamish, you would know that you could grow hemp inside of a forest quite well. It grows quite well.

Hamish Kimmins

I'd like to talk to you about that.

Steve Smith

I'm not familiar with any of the economic analysis either, quite frankly, but if the economics would support complete replacement of wood fibers with straw fibers, my guess is that inevitably, it will happen.

Peter Etheridge

Okay, we have run out of time, and with that I would like to thank the panel for completing the task here today, and you the audience and the people who have made comments for producing a fruitful discussion. I would like you to give a round of applause to the panel. Thank you.

Index

Page numbers followed by "f" indicate figures; page numbers followed by "t" indicate tables.

357

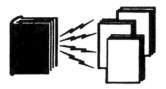

Haworth
DOCUMENT DELIVERY
SERVICE

This valuable service provides a single-article order form for any article from a Haworth journal.

- *Time Saving:* No running around from library to library to find a specific article.
- *Cost Effective:* All costs are kept down to a minimum.
- *Fast Delivery:* Choose from several options, including same-day FAX.
- *No Copyright Hassles:* You will be supplied by the original publisher.
- *Easy Payment:* Choose from several easy payment methods.

Open Accounts Welcome for ...
- Library Interlibrary Loan Departments
- Library Network/Consortia Wishing to Provide Single-Article Services
- Indexing/Abstracting Services with Single Article Provision Services
- Document Provision Brokers and Freelance Information Service Providers

---

MAIL or *FAX* THIS ENTIRE ORDER FORM TO:

Haworth Document Delivery Service
The Haworth Press, Inc.
10 Alice Street
Binghamton, NY 13904-1580

or FAX: 1-800-895-0582
or CALL: 1-800-342-9678
9am-5pm EST

PLEASE SEND ME PHOTOCOPIES OF THE FOLLOWING SINGLE ARTICLES:

1) Journal Title: _____

   Vol/Issue/Year: _____ Starting & Ending Pages: _____

Article Title: _____

_____

2) Journal Title: _____

   Vol/Issue/Year: _____ Starting & Ending Pages: _____

Article Title: _____

_____

3) Journal Title: _____

   Vol/Issue/Year: _____ Starting & Ending Pages: _____

Article Title: _____

_____

4) Journal Title: _____

   Vol/Issue/Year: _____ Starting & Ending Pages: _____

Article Title: _____

(See other side for Costs and Payment Information)

COSTS: Please figure your cost to order quality copies of an article.

1. Set-up charge per article: $8.00
   ($8.00 × number of separate articles) _____

2. Photocopying charge for each article:

   1-10 pages: $1.00 _____

   11-19 pages: $3.00 _____

   20-29 pages: $5.00 _____

   30+ pages: $2.00/10 pages _____

3. Flexicover (optional): $2.00/article _____

4. Postage & Handling: US: $1.00 for the first article/
   $.50 each additional article _____

   Federal Express: $25.00 _____

   Outside US: $2.00 for first article/
   $.50 each additional article _____

5. Same-day FAX service: $.35 per page _____

GRAND TOTAL: _____

METHOD OF PAYMENT: (please check one)

❑ Check enclosed ❑ Please ship and bill. PO # _____
(sorry we can ship and bill to bookstores only! All others must pre-pay)

❑ Charge to my credit card: ❑ Visa; ❑ MasterCard; ❑ Discover;
❑ American Express;

Account Number:_____ Expiration date:_____

Signature: ✗_____

Name: _____ Institution: _____

Address: _____

_____

City: _____ State:_____ Zip:_____

Phone Number: _____ FAX Number: _____

MAIL or *FAX* THIS ENTIRE ORDER FORM TO:

Haworth Document Delivery Service	**or FAX:** 1-800-895-0582
The Haworth Press, Inc.	**or CALL:** 1-800-342-9678
10 Alice Street	9am-5pm EST)
Binghamton, NY 13904-1580	